Nuclear Power
Is Not the Answer

Also by Helen Caldicott
from The New Press

The New Nuclear Danger:
George W. Bush's Military-Industrial Complex

Nuclear Power
Is Not the Answer

Helen Caldicott

THE NEW PRESS

NEW YORK
LONDON

For my beloved grandchildren
Mikhael, Rachel, Paul, Claire, Oscar and Tess.

Published in the United States by The New Press, New York, 2006
Distributed by W. W. Norton & Company, Inc., New York

LIBRARY OF CONGRESS CATALOGING-IN-PUBLICATION DATA

Caldicott, Helen.
    Nuclear power is not the answer / Helen Caldicott.
        p. cm.
    Includes bibliographical references and index.
    ISBN 978-1-59558-067-2 (hc)
    ISBN 978-1-59558-213-3(pbk)
        1. Nuclear industry—United States.   2. Nuclear energy—United States.
3. Global warming—Prevention.   4. Nuclear power plants—United States.   5. Energy
policy—United States.   I. Title.

HD9698.U52C33   2006
333.792'40973—dc22   2006041990

The New Press was established in 1990 as a not-for-profit alternative to the large, commercial publishing houses currently dominating the book publishing industry. The New Press operates in the public interest rather than for private gain and is committed to publishing, in innovative ways, works of educational, cultural, and community value that are often deemed insufficiently profitable.

www.thenewpress.com

Composition by Westchester Book Group

Printed in the United States of America

10 9 8 7 6 5 4 3 2 1

Contents

Acknowledgments

I have much gratitude for the invaluable help, knowledge, advice, and support provided by the many people who assisted in compiling this book. They include Mary Cunnane, Jan Willem Storm van Leeuwen, Philip Smith, Tom Cochran, Dan Hirsch, Arjun Makhijani, Brice Smith, Paul Craig, David Lochbaum, Gordon Thompson, Bob Alveraz, Paul Gunter, Anna Aurilio, Kay Drey, Ed Lyman, David Richardson, Steve Wing, Amory Lovins, Harvey Wasserman, Tom Fasy, Kevin Kamps, Charles Sheehan-Miles, Bruce Biewald, Mary Osborn, Scott Powell, Ernest Sternglass, and Peter Bradford.

Last but not least I would like to thank my editor, Diane Wachtell, who conceived and guided this book to fruition.

Introduction

"[Nuclear power] is a very important part of our energy policy today in the U.S. . . . America's electricity is already being provided through the nuclear industry efficiently, safely, and with no discharge of greenhouse gases or emissions."

—Vice President Cheney in a speech
to the Nuclear Energy Institute, May 22, 2001

"The 103 nuclear power plants in America produce 20% of the nation's electricity without producing a single pound of air pollution or greenhouse gases."[1]

—President Bush in a speech to a group
of nuclear power plant workers at the
Calvert Cliffs nuclear reactor, June 22, 2005

The current administration clearly believes that if it lies frequently and with conviction, the general public will be lulled into believing their oft-repeated dictums. As this book will show, no part of "efficiently, safely, and with no discharge of greenhouse gases or emissions" is true. Nuclear energy creates significant greenhouse gases and pollution today, and is on a trajectory to produce as much as conventional sources of energy within the next one or two decades. It requires massive infusions of government (read taxpayer) subsidies, relying on universities and the weapons industry for its research and development, and being considered far too risky for private investors. It is also doubtful that the 8,358 individuals diagnosed between 1986 and 2001 with thyroid cancer in Belarus, downwind of Chernobyl, would choose the adjective "safe" to describe nuclear power.

Nuclear power is not "clean and green," as the industry claims, because large amounts of traditional fossil fuels are required to mine and refine the uranium needed to run nuclear power reactors, to construct the massive concrete reactor buildings, and to transport and store the toxic radioactive waste created by the nuclear process. Burning of this fossil fuel emits significant quantities of carbon dioxide (CO_2)—the primary "greenhouse gas"—into the atmosphere. In addition, large amounts of the now-banned chlorofluorocarbon gas (CFC) are emitted during the enrichment of uranium. CFC gas is not only 10,000 to 20,000 times more efficient as an atmospheric heat trapper ("greenhouse gas") than CO_2, but it is a classic "pollutant" and a potent destroyer of the ozone layer.

While currently the creation of nuclear electricity produces only one-third the amount of CO_2 emitted from a similar-sized, conventional gas generator, this is a transitory statistic. Over several decades, as the concentration of available uranium ore declines, more fossil fuels will be required to extract the ore from less-concentrated ore veins. Within ten to twenty years, nuclear reactors will produce no net energy because of the massive amounts of fossil fuel that will be necessary to mine and to enrich the remaining poor grades of uranium. (The nuclear power industry contends that large quantities of uranium can be obtained by reprocessing radioactive spent fuel. However, this process is extremely expensive, medically dangerous for nuclear workers, and releases large amounts of radioactive material into the air and water; it is therefore not a pragmatic consideration.) By extension, the operation of nuclear power plants will then produce exactly the same amounts of greenhouse gases and air pollution as standard power plants.

Contrary to the nuclear industry claims, smoothly running nuclear power plants are also not emission free. Government regulations allow nuclear plants "routinely" to emit hundreds of thousands of curies of radioactive gases and other radioactive elements into the environment every year. Thousands of tons of solid radioactive waste are presently accumulating in the cooling pools beside the 103 operating nuclear plants in the United States and hundreds of others throughout the world. This waste contains extremely toxic elements that will inevitably pollute the environment and human food chains, a legacy that will lead to epidemics of cancer, leukemia, and genetic disease in populations living near nuclear power plants or radioactive waste facilities for many generations to come.

Nuclear power is exorbitantly expensive, and notoriously unreliable. Wall Street is deeply reluctant to re-involve itself in any nuclear investment, despite the fact that in the 2005 Energy Bill the U.S. Congress allocated $13 billion in subsidies to revive a moribund nuclear power industry. To compound this problem, the global supplies of usable uranium fuel are finite. If the entire world's electricity production were replaced today by nuclear energy, there would be less than nine more years of accessible uranium. But even if certain corporate interests are convinced that nuclear power at the moment might be a beneficial investment, one major accident at a nuclear reactor that induces a meltdown would destroy all such investments and signal the end of nuclear power forever.

In this day and age, nuclear power plants are also obvious targets for terrorists, inviting assault by plane, truck bombs, armed attack, or covert intrusion into the reactor's control room. The subsequent meltdown could induce the death of hundreds of thousands of people in heavily populated areas, and they would expire slowly and painfully, some over days and others over years from acute radiation illness, cancer, leukemia, congenital deformities, or genetic disease. Such an attack at the Indian Point reactors, thirty-five miles from Manhattan, for instance, would effectively incapacitate the world's main financial center for the rest of time. An attack on one of the thirteen reactors[2] surrounding Chicago would wreak similar catastrophic medical consequences. Amazingly, security at U.S. nuclear power plants remains at virtually the same lax levels as prior to the 9/11 attacks.

Adding to the danger, nuclear power plants are essentially atomic bomb factories. A 1,000 megawatt nuclear reactor manufactures 500 pounds of plutonium a year; normally ten pounds of plutonium is fuel for an atomic bomb. A crude atomic bomb sufficient to devastate a city could certainly be crafted from reactor grade plutonium. Therefore any non-nuclear weapons country that acquires a nuclear power plant will be provided with the ability to make atomic bombs (precisely the issue the world confronts with Iran today). As the global nuclear industry pushes its nefarious wares upon developing countries with the patent lie about "preventing global warming," collateral consequences will include the proliferation of nuclear weapons, a situation that will further destabilize an already unstable world.

Meanwhile, every billion dollars spent on the supremely misguided attempt to revivify the nuclear industry is a theft from the production of cheap renewable electricity. Think what these billions could do if invested in the development of wind power, solar power, cogeneration, geothermal energy, biomass, and tidal and wave power, let alone basic energy conservation, which itself could save the United States 20% of the electricity it currently consumes.

A Greenpeace report issued in October 2005 predicted that solar power could supply clean electricity to 100 million people living in sunny parts of the world by the year 2025. Such an enterprise could create 54,000 jobs and be worth $19.9 billion. In just two decades, the amount of solar electricity could be equivalent to the power generated by seventy-two coal-fired stations—for example, enough to supply the needs of Israel, Morocco, Algeria, and Tunisia combined. (Egypt is currently one of the few countries in the world that hosts a government department solely devoted to the development of renewable energy sources.[3])

The Carbon Trust, an independent company established by the British government, estimates that, with the correct amount of investment, marine energy—tidal and wave power—could provide up to 20% of the United Kingdom's current electricity needs. As Marcus Rand, chief executive of the British Wind Energy Association, said, "The report provides impetus behind the vision that Britain can rule the waves and the tides making a significant dent in our carbon emissions alongside creating new world-class industries for the UK."[4]

According to Amory Lovins, CEO of the Rocky Mountain Institute, in 2004 the amount of electricity supplied by renewable energy sources—wind, co-generation, biomass, geothermal, solar, hydro (excluding electricity generated from large hydro dams)—added 5.9 times the total capacity worldwide that nuclear power contributed, and raised the global electricity production 2.9 times more than nuclear power contributed. These "minor" electricity sources already dwarf the annual growth of nuclear power generation, and experts predict that by 2010, they will add 177 times more capacity than nuclear power provides.[5]

When nuclear proponents say that nuclear power can be used to reduce the United States's insatiable reliance on foreign oil, they are simply wrong. Oil and its by-product gasoline are used to fuel the internal combustion engines in automobiles and trucks. Oil is also used to heat buildings. But oil does not power the electric grid. The grid, which is used to power electric lights, computers, VCRs, fans, hair dryers, stoves, refrigerators, air conditioners, and for industrial needs, is powered primarily through the burning of coal, other fossil fuels, and, currently, through nuclear power. (Oil does generate an infinitesimal amount of electricity—2% in the United States.)

How exactly is electricity generated? In the case of hydropower (which accounts for 7% of the electricity generated in the United

States) the momentum of falling water is converted into electricity. For most of the remaining 93%, coal (50%), natural gas (18%), nuclear power (20%), and oil (2%) are used to produce immense amounts of heat. The heat boils water, converting it to steam, which then turns a turbine, generating electricity. So, in essence, a nuclear reactor is just a very sophisticated and dangerous way to boil water—analogous to cutting a pound of butter with a chain saw. At the moment, hydro provides 7%, and unfortunately wind is only 2% of the total U.S. mix, while solar is less than 1%. Globally, coal supplies about 64% of the world's electricity, hydro and nuclear each provide 17%, and renewable sources again make up 2%.[6]

Tragically, more and more people are believing the myths propagated by the nuclear industry about nuclear power—that it is emission free, produces no greenhouse gases, and is therefore the answer to global warming. Before the British election in May 2005, the nuclear industry slowly and surely fashioned a classy public relations campaign targeting politicians, media, and the British public. (That campaign, coordinated by the Nuclear Industry Association, cleverly did not address the dubious benefits of nuclear power but focused instead upon the current shortcomings of wind-generated electricity and other alternative power sources.[7])

The British Department of Trade and Industry (DTI) also viewed the 2005 election as an opportunity to promote nuclear power. Adrian Gault, director of DTI's strategy group, made a wild and uninformed prediction that nuclear power would be supplying half of Britain's electricity by 2050 while cutting greenhouse emissions. (Meanwhile, in 2001, DTI's Nuclear Industries Directorate had already agreed to participate in an international consortium to build the next generation of nuclear reactors—to be constructed by a British or American company. So their real agenda had been established four years earlier, and the propaganda campaign in May 2005 was merely an attempt to bring the British public around to seeing the wisdom of preordained policy.[8])

The British nuclear industry is working hard to persuade members of parliament and other influential public figures of the benefits of nuclear power. Dr. James Lovelock, the UK-based scientist who developed the Gaia theory, now wrongly advocates the use of nuclear power as one solution to the global warming crisis.[9] Sir David King, chief UK government science advisor, says that nuclear power plants are the only realistic way to satisfy growing energy demands while meeting global warming targets.[10] The British nuclear industry has sacrificed full disclosure and jettisoned truth in order to ensure a new round of government subsidies for nuclear power. The government subsidy program for the nuclear industry—which might be dubbed the "Security of Supply Obligation"—amounts in essence to the socialization of nuclear power, ensconced within a "free market" economy.[11]

In England in 2006, nuclear power has risen to the top of the political agenda, as government ministers and public officials rush to address an impending energy crisis, driven by Russia's January 2006 decision to cut off its natural gas supplies to the Ukraine and hence to much of Europe. This scare helped to convince an already compliant Prime Minister Blair and senior people at the UK Department of Trade and Industry that new nuclear power stations are needed.[12]

In the United States and Canada, leading environmentalists similarly seem to have been swayed by the Bush/Cheney/nuclear industry rhetoric. Stewart Brand, founder of the Whole Earth Catalogue;[13] Gus Speth, the dean of Yale's School of Forestry and Environmental Studies;[14] and former Greenpeace Canada leader Patrick Moore, who now consults for the mining, fishing, and timber industries[15]—all seem to have accepted the nuclear industry's propaganda as fact. Meanwhile, it is increasingly critical to set the record on nuclear power straight, as international battles for oil threaten to morph into world wars, and leading NASA scientists are taken to task by the Bush administration for daring to tell the truth about global warming.[16]

It is interesting to speculate why President Bush and Vice President Cheney are so beholden to and enamored of the nuclear power industry, an industry that has never actually been exposed to the chill winds of the market economy they unfailingly espouse elsewhere. As neither the president nor the vice president can boast a scientific education, they would be hard pressed to understand the scientific and medical problems associated with this arcane industry.[17] Both are oil men who have made a great deal of money directly or indirectly through that industry; they are deeply indebted to big business for political contributions; and they overtly seem not to be interested in the health and well being of the American people, let alone the dire situation facing the planet in the form of global warming, and the threat of nuclear meltdowns and nuclear pollution.

Ironically, while the Bush administration is reluctant to admit that global warming is really happening and that it could be caused by deleterious human activities, it is using the issue of global warming to justify the increased production of nuclear power, which, it claims, is the answer to (the non-existent problem of) global warming. Claiming, as Cheney does, that atomic electricity produces no carbon dioxide, the culprit responsible for 50% of atmospheric heating,[18] the U.S. nuclear propaganda apparatus has been shifted into high gear to convince politicians and public alike that there can be and will be no other reasonable solution apart from nuclear power to answer this catastrophic global problem now threatening many life forms with extinction. Global warming has been a great gift to the nuclear industry.

Fewer than ten days after taking office, Cheney promised to "restore decency and integrity to the oval office," while he simultaneously took charge of the administration's energy task force, called the National Energy Policy Development Group.[19] On April 17, 2001, Cheney met with Kenneth Lay, the CEO of the now disgraced Enron Corporation, to discuss "energy policy matters" and the "energy crisis in California." Following that meeting, Lay gave

Cheney a three-page wish list of corporate recommendations. A subsequent comparison of that memo against the final report of the National Energy Policy Development Group showed that the task force had adopted all or significant portions of the Lay memo in seven of eight policy areas. In total, seventeen policies sought by Enron were adopted.[20]

Cheney and his aides met at least six times with Lay and other Enron officials while preparing the task force report, which is now the basis of the administration's energy proposals. Cheney's staff also met with an Enron sponsored lobbying organization, the "Clean Power Group." Cheney, his aides, and cabinet departments have repeatedly refused requests for the records of these meetings, despite the fact that the Federal Advisory Committee Act of 1972 says that task forces like Cheney's must conduct public meetings and must keep publicly available records.[21] While we do not know, as a result, what Enron may have advocated in that meeting with respect to nuclear energy, we do know that Enron made significant contributions to the Bush/Cheney campaign, the Florida recount fight fund, and to the Bush/Cheney inauguration—a situation that calls into question whether legal and ethical guidelines were crossed.[22]

The American Nuclear Society recently held a meeting in San Diego that drew scientists and industry professional from all around the world. The prevailing mantra was simple—surprise the opponent, plan ahead, coordinate, be pro-active not reactive, and engage and communicate with antinuclear groups.[23] This extensive propaganda campaign is global. A formally chartered organization composed of the governments of Argentina, Brazil, Canada, the European Union, France, Japan, the Republic of Korea, the Republic of South Africa, Switzerland, the United Kingdom, and the United States, called the Generation IV International Forum (GIF), is collaborating with the U.S. Nuclear Energy research Advisory Committee to elucidate the benefits, technical and institutional barriers, and research needs for the most promising nuclear energy system concepts.

Other countries engaged in the possible construction of nuclear power plants include China, which already has nine nuclear reactors and plans to build another thirty nuclear power plants. (Even if it builds its thirty plants, however, nuclear power will still provide only 5% of its electricity mix, while the percentage of China's electrical generation capacity by natural gas is expected to increase from 1% today to over 6% by 2030 according to the International Energy Agency.[24]) New nuclear power capacity is under consideration or construction in India, Japan, Taiwan, Turkey, Belarus, Vietnam, Poland, and South Korea. Russia as well as Finland have several plants under construction.[25]

Nuclear power is often referred to behind closed doors in the U.S. Department of Energy as "hard" energy whereas wind power, solar power, hydropower, and geothermal energy are referred to as "soft" energy pathways. Clearly the same psychosexual language used by the Pentagon generals to describe various aspects of nuclear weapons and nuclear war has been translocated into the nuclear power vocabulary of some very powerful and influential men in the electricity generating field.[26] As a physician, I contend that unless the root cause of a problem can be ascertained there can be no cure. So too the pathology intrinsic in the nuclear power gang needs to be dissected and revealed to the cold light of day.

The potential for growth in the renewable non-CO_2 producing sectors is enormous. All that is required is a commitment by government leaders to urgently enact serious laws mandating energy conservation, and to shift the subsidies currently provided to the nuclear power industry to alternative and renewable electricity generation. Corporations as well should be given incentives to invest in exciting and diverse non-polluting energy technologies. In truth, the earth is in the intensive care unit, and the prognosis is poor indeed unless we all take courageous measures.

Nuclear Power
Is Not the Answer

The Energetic Costs of Nuclear Power

The Nuclear Energy Institute (NEI), the propaganda wing and trade group for the American nuclear industry, spends millions of dollars annually to engineer public opinion. Advertisements such as the one on page 5 have been published extensively by the NEI in *Scientific American*, the *New Yorker*, the *Washington Post*, and Capitol Hill publications such as *Roll Call*, *Congress Daily AM*, and *The Hill*.[1] The primary goal of such ads is to establish the premise that nuclear energy is "cleaner and greener" than traditional sources of electricity. Sentences such as "our 103 nuclear power plants don't burn anything, so they don't produce greenhouse gases" imply that nuclear energy is a more environmentally conscious choice than, say, electricity produced from coal or oil—the traditional sources of fuel across the globe—one that will produce far less carbon dioxide and thus spare us the global warming problems now associated with these other energy sources.

But a clear-eyed look at the true costs of nuclear energy production tells a very different story. The fact is, it takes energy to make energy—even nuclear energy. And the true "energetic costs" of making nuclear energy—the amounts of traditionally generated fuel it takes to create "new" nuclear energy—have not been tallied up until very recently. Certainly, they are absent from the NEI ads.

What exactly is nuclear power? It is a very expensive, sophisticated, and dangerous way to boil water. Uranium fuel rods are placed in water in a reactor core, they reach critical mass, and they produce vast quantities of heat, which boils the water. Steam is directed through pipes to turn a turbine, which generates electricity. The scientists who were involved in the Manhattan Project creating nuclear weapons developed a way to harness nuclear energy to generate electricity. Because their guilt was so great, they were determined to use their ghastly new invention to help the human race.[2] Nuclear fission harnessed "atoms for peace," and the nuclear PR industry proclaimed that nuclear power would provide an endless supply of electricity—referred to as "sunshine units"—that would be good for the environment and "too cheap to meter."

They were wrong. Although a nuclear power plant itself releases no carbon dioxide, the production of nuclear electricity depends upon a vast, complex, and hidden industrial infrastructure that is never featured by the nuclear industry in its propaganda, but that actually releases a large amount of carbon dioxide as well as other global warming gases. One is led to believe that the nuclear reactor stands alone, an autonomous creator of energy. In fact, the vast infrastructure necessary to create nuclear energy, called the nuclear fuel cycle, is a prodigious user of fossil fuel and coal.

The production of carbon dioxide (CO_2) is one measurement that indicates the amount of energy used in the production of the nuclear fuel cycle. Most of the energy used to create nuclear energy—to mine uranium ore for fuel, to crush and mill the ore, to enrich the uranium, to create the concrete and steel for the reactor, and to store the thermally and radioactively hot nuclear waste—comes from the consumption of fossil fuels, that is, coal or oil. When these materials are burned to produce energy, they form CO_2 (reflecting coal and oil's origins in ancient trees and other organic carboniferous material laid down under the earth's crust

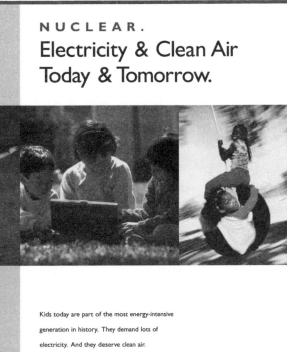

NUCLEAR.
Electricity & Clean Air Today & Tomorrow.

Kids today are part of the most energy-intensive generation in history. They demand lots of electricity. And they deserve clean air.

That's why nuclear energy is so important to America's energy future. Nuclear energy already produces electricity for 1 of every 5 homes and businesses. And our 103 nuclear power plants are emission free so they help keep the air clean.

We need secure, reliable sources of electricity for the 21st Century—and we also need clean air. With nuclear energy, we can have both.

NUCLEAR.
CLEAN AIR ENERGY.

Nuclear energy is the most reliable source of electricity.

WWW.NEI.ORG

A TYPICALLY FALLACIOUS AND MISLEADING NUCLEAR
ENERGY ADVERTISEMENT

millions of years ago). For each ton of carbon burned, 3.7 tons of CO_2 gas are added to the atmosphere, and this is the source of to-day's global warming.

CO_2 and other gases hover in the lower atmosphere or tropo-sphere, covering the earth like a blanket, and this gaseous layer be-haves like glass in a greenhouse. Visible white light from the sun enters the atmosphere, heating up the surface of the earth, but the infrared heat radiation created cannot pass back through the terres-trial layer of trapped gases. Carbon dioxide accounts for 50% of the global warming phenomenon,[3] and other rare gases comprise the rest.[4]

The total energy input of the nuclear fuel cycle—the ener-getic costs of nuclear power—must be openly and honestly assessed if nuclear power is to be compared fairly with other energy sources. Very few studies are yet available that analyze the total life cycle of nuclear power and its final energy input versus output. One of the best is a study by Jan Willem Storm van Leeuwen and Philip Smith titled "Nuclear Power—the Energy Balance." Much of the material for the next section has been derived from this ex-cellent report.

To quote the final conclusion of their lengthy analysis, "The use of nuclear power causes, at the end of the road and under the most favourable conditions, approximately one-third as much car-bon dioxide (CO_2) emission as gas-fired electricity production. The rich uranium ores required to achieve this reduction are, however, so limited that if the entire present world electricity demand were to be provided by nuclear power, these ores would be exhausted within nine years. Use of the remaining poorer ores in nuclear re-actors would produce more CO_2 emission than burning fossil fuels directly."[5] In this instance, nuclear reactors are best understood as complicated, expensive, and inefficient gas burners.[6]

The nuclear fuel cycle is composed of many interesting and complicated steps, each of which entails its own energetic costs.

The next sections enumerate the parts of the nuclear fuel cycle and examine the energy input necessary for each step. (These energetic analyses are rough estimates, but they are the best available at this time.)

URANIUM MINING AND MILLING

The largest unavoidable energy cost associated with nuclear power relates to the processes of mining and milling uranium fuel. Variable grades of uranium ore exist at different mines around the world. A greater amount of energy is required to extract uranium from a mine containing a low-grade uranium concentration of 0.1% than from another mine containing a uranium concentration of 1%— ten times more. Therefore the specific energy expenditure required for uranium extraction from the original ore body is largely dependent upon the ore grade. The energy used to mine the uranium is fossil fuel—the kind of energy nuclear power is touted as replacing—with the concurrent production of carbon dioxide.

There is a point at which the concentration of uranium becomes so low that the energy required to extract and to refine a dilute uranium ore concentration from the ground is greater than the amount of electricity generated by the nuclear reactor. For example, 162 tons of natural uranium must be extracted from the earth's crust each year to fuel one nuclear power plant. If the uranium is in granite ore, with a low-grade uranium concentration of 4 grams per ton of rock (0.0004%), then 40 million tons of granite will need to be mined. This rock will need to be ground into fine powder and chemically treated with sulphuric acid and other chemicals to extract the uranium from the rock (milling). Assuming an extraction capacity of 50% (an unrealistically high estimate), 80 million tons of granite will therefore need to be treated. The dimensions of this mass of rock are one hundred meters high and three kilometers long. The extraction of uranium from this granite rock would consume over thirty times the energy generated in the reactor from the extracted uranium.[7]

The high-grade uranium ores are finite—global high-grade reserves amount to 3.5 million tons. Given that the current use of uranium is about 67,000 tons per year, these reserves would supply fifty more years of nuclear power at current production levels (but only nine years, as noted above, if all the world's electricity needs were met by nuclear energy). The total of all the uranium reserves, including high and low grade, is estimated to be approximately 14.4 million tons, but most of these ores would be extremely expensive to mine, and the ore grades would be too low for electricity production. Many uranium mines are therefore out of use already.[8]

The mining and milling of uranium is a complex process. The rock itself must be excavated by bulldozers and shovels and then transported by truck to the milling plants. All these machines use diesel oil. Furthermore, the maintenance shops that service this equipment consume electricity and hence fuel oils. The uranium-bearing rock is then ground to a powder in electrically powered mills; the powder is treated with chemicals, usually sulphuric acid; then several other chemicals (many of which are highly corrosive and poisonous) are used to convert the uranium to a compound called yellow cake. Fuel is also needed during this process to create steam and heated gases, and all the chemicals used in the mills must be manufactured at other chemical plants.

The specific energy expenditure of the milling process depends upon which of the two types of available ore are processed. Soft ores, in which uranium is contained in sandstones, shales, and calcretes, with uranium concentrations ranging from 10% down to 0.01%, require 2.33 gigajoules per ton of ore extracted (1 gigajoule = 1 billion joules).[9] Hard ores, including quartz pebble conglomerates and granites, with grades that vary from 0.1% to 0.001% or less, require 5.5 gigajoules per ton of ore extracted. In either case, when the ore grade reaches 0.01% the nuclear fuel cycle becomes energetically non-productive, because so much energy is expended to mine and mill the low-grade ores.[10]

MILL TAILINGS

If the mill tailings that remain after the extraction of the uranium were to be subject to remediation, as they should be, massive quantities of fossil fuel would be required for this process as well. Millions of tons of radioactive material that is currently dumped on the ground, often on native Indian tribal land, emitting radioactive elements to the air and water, need instead to be buried deeply in the ground where the uranium originally emanated. This single remediation process, which should be scrupulously observed, by itself makes the energetic price of nuclear electricity unreasonable.[11]

These tailings would need to be:

- neutralized with limestone;
- immobilized by mixing them with bentonite to isolate them from ground water;
- transported and placed back into the mine;
- covered with overburden or soil and then with indigenous vegetation.

The energy expenditure for adequate remediation is estimated to be 4.2 gigajoules per metric ton of tailings, four times the 1.06 gigajoules per metric ton expended on the original mining. The remediation process also involves the extensive use of fossil fuels and the production of more carbon dioxide.[12]

CONVERSION OF URANIUM TO URANIUM HEXAFLUORIDE

Before uranium can be enriched, it must be converted to uranium hexafluoride gas, because it is in this form that the fissionable uranium 235 can be separated from the non-fissionable uranium 238. Uranium hexafluoride is the only uranium compound that is gaseous at low temperatures and therefore is easy to work with. The specific energetic requirements for this conversion are 1.478 gigajoules per kilogram of uranium.

URANIUM ENRICHMENT

Enrichment of uranium 235 from 0.7% to 3% is also a very energy-consuming process. Specific energy expenditures for enrichment include construction, operation, and maintenance of the enrichment plant. Uranium can be enriched using one of two basic methods—gaseous diffusion and ultracentrifuge—both of which require very large amounts of energy. (Enrichment by ultracentrifuge has a lower direct energy cost, but the financial costs of the operation and maintenance of ultracentrifuge enrichment are much higher than gaseous diffusion because of the short technical life of the centrifuges.)

In the United States, enrichment facilities have historically been located at Paducah, Kentucky, and Portsmouth, Ohio, with a discarded facility at Oak Ridge, Tennessee. In 2001, however, the privately owned and operated United States Enrichment Corp. consolidated its operation in Paducah. The Paducah enrichment facility uses the electrical output of two dirty, old 1,000 megawatt coal-fired plants for its operation,[13] contributing significant carbon dioxide to the atmosphere. It has also recently been revealed by the U.S. Department of Energy that CFC 114 gas—a compound that is a potent global warmer and that destroys the stratospheric ozone layer—leaks unabated from the hundreds of miles of cooling pipes used in the uranium enrichment operation at Paducah, Kentucky, and its sister facility in Ohio.[14]

The specific energetic costs of enrichment are measured in joules per separative work unit (SWU). Averaging the current world use of the two different processes—30% gaseous diffusion and 70% ultracentrifuge—the energetic costs are 0.0055 petajoules per 1,000 SWU. (A petajoule is 1 million billion joules.)[15]

FUEL ELEMENT FABRICATION

The enriched uranium hexafluoride gas is then made into solid fuel pellets of uranium dioxide, the size of a cigarette filter. These uranium pellets are put into zirconium fuel rods which are twelve feet long and half an inch thick. A typical 1,000 megawatt reactor contains 50,000 of these fuel rods—about one hundred tons of uranium. Again fossil fuel is used in the fabrication process, and the specific energy expenditure is 0.00379 petajoules per ton of uranium.

CONSTRUCTION OF THE REACTOR

All nuclear power plants in the United States were constructed between the years 1980 to 1985 or before, and no new plants have been ordered since 1978. The construction of a nuclear power plant requires an immense aggregate of goods and services. Nuclear technology is a very high-tech process, requiring an extensive industrial and economic infrastructure. A huge amount of concrete and steel is used to build a reactor. Furthermore, construction has become ever more complex because of increased safety concerns following the meltdowns at Three Mile Island and Chernobyl.

Estimates vary for the energetic costs of reactor construction from 40 to 120 petajoules. The mean value of 80 petajoules has been used in the study of Storm van Leeuwen and Smith.

DECOMMISSIONING AND DISMANTLING

When the reactor is finally closed at the end of its working life, the intensely radioactive products—cobalt 60 and iron 55 formed inside the reactor vessel from neutron bombardment—must be allowed to decay considerably before the reactor can even be entered. (Additional residual contaminating radioactive elements, which are also very dangerous, include tritium, carbon 14, and calcium 41, among others.[16]) Thus, these huge, intensely radioactive mausoleums must be guarded and protected from damage or unwarranted intrusion for a period of ten to hundred years before the actual process of dismantling can begin.

The steps involved in decommissioning and dismantling include:

- operation and maintenance of the reactor during the safe-guarded period after the final shutdown;
- clean-up of the radioactive parts of the reactor before dismantling;
- demolition of the radioactive components;
- dismantling;
- packaging and permanent disposal of the dismantled wastes.

After sufficient time is given for the radioactive decay period, the reactor must be cut apart into small pieces either by humans or by remote control, and the still-radioactive pieces must be packed into containers for transportation and final disposal at some distant location. There is very limited experience available on which to base energetic cost estimates for decommissioning and dismantling, because a large nuclear power plant has never actually been dismantled completely after a long operational lifetime. However, based on the scarce available data, the energetic debt for this exercise is estimated to be in the range of 80–160 petajoules, the high end of the range being the most probable.[17] Traditional coal- or gas-fired plants can be dismantled in the conventional way as any building, because they are not radioactive and therefore do not pose a risk to the public health and safety. The discarded materials, rubble, and scrap from conventional buildings can be reused. For comparison: Construction and dismantling of a gas-fired plant require about 24 petajoules together. The energy requirements of construction and dismantling of a nuclear power plant may sum up to about 240 petajoules.

CLEANUP

At the end of its lifetime, the reactor will need to be cleaned of extensive quantities of accumulated radioactive material called CRUD (Chalk River Unidentified Deposits, so named because these materials were first found in the Chalk River reactor). CRUD is a collection of radioactive elements that come from the reactor itself—from the cooling system and the highly radioactive fission and "actinide" elements that have escaped from leaking and damaged fuel rods. This process, which is separate from decommissioning, may be energetically very expensive and will need as much energy debt as 50% of the original energetic construction costs, which is 20 to 60 petajoules.[18]

COOLING WATER: TRITIUM AND CARBON 14

The water that cools the reactor core becomes heavily contaminated with tritium, or radioactive hydrogen, and with carbon 14, the long-term medical and ecological effects of which are not well understood and are rarely discussed or addressed by the nuclear industry or anyone else. The radioactive life of tritium is more than 200 years, and the radioactive life of carbon 14 is 114,600 years. A sustainable energy system would necessitate a closed loop for tritium and carbon 14, such that they never enter the ecosphere. Theoretically this water should be stored, immobilized into drying agents or into cement, and placed in appropriate long-lived containers. Instead, it is routinely and blithely discharged into seas, rivers, or lakes, from which people obtain their drinking water.[19] Implementing proper disposal techniques would require a huge number of waste containers and massive energy expenditure.

The fact that there is thus far no adequate knowledge of the long-term biological dangers and because of the absolutely immense expense associated with sequestering the tritium and carbon 14 from nuclear power plants, there is no adequate estimate of the energetic costs required to prevent the release of these isotopes. Hence, the true energetic and economic costs of nuclear power are presently grossly underestimated.[20]

DISPOSAL OF RADIOACTIVE WASTE

Radioactive waste is classified in vaguely defined categories as low level, intermediate level, and high level, according to the concentration and types of radioactive elements. There are five types of specific containers available to transport these wastes depending upon the category, which are labelled V1 to V5. The production, filling, handling, and transport of the radioactive waste in containers V2 to V4 is estimated to use per ton almost as much energy as the specific construction energy of the atomic power reactor itself. The total may sum up to a very large amount as noted previously—about 20 petajoules.[21]

In addition to handling the reactor wastes, the energetic costs of nuclear electricity include those associated with interim storage of irradiated fuel elements. The magnitude of the radiation generated in a nuclear power plant is almost beyond belief. The original uranium fuel that is subject to the fission process becomes 1 billion times more radioactive in the reactor core.[22] A thousand megawatt nuclear power plant contains as much long-lived radiation as that produced by the explosion of one thousand Hiroshima-sized bombs.

Every year, one-third of the now-intensely radioactive fuel rods must be removed from the reactor, because they are so contaminated with fission products that they hinder the efficiency of the electricity production.

These rods emit so much radiation that a lethal dose can be acquired by a person standing in close proximity to a single spent fuel rod within seconds. But they are also extremely thermally hot and must therefore be stored for thirty to sixty years in a heavily shielded building and continually cooled by air or water. If they are not continually cooled, the zirconium cladding of the rod could become so hot that it would spontaneously burn, releasing its radioactive inventory. Finally, after an adequate cooling period, the rods must eventually be packed into a container by remote control.

Construction of these highly specialized containers uses as much energy as construction of the original reactor itself, which is 80 gigajoules per metric ton. To make matters worse, spent fuel packaging is a completely new and relatively untested technology for which there is no operational data.[23]

TRANSPORTATION OF HIGH-LEVEL AND INTERMEDIATE WASTE AND LONG-TERM STORAGE FOR 240,000 YEARS

The calculations for this part of the nuclear fuel cycle have not yet been done. But clearly, huge amounts of fossil fuel will be used to transport the waste over long distances through many towns and cities over long periods of time, to prepare an adequate geological waste storage facility, and to supervise and guard the site for periods of time almost beyond our comprehension—240,000 years.[24]

Energetic cost assessments provided by the global energy industry are notoriously and consistently fallacious. For instance, BP-Amoco in its 2005 world energy supply assessment, simplistically assessed only the gross electricity production of nuclear power plants, but failed to incorporate the total energy consumption of the nuclear fuel chain.[25]

In fact, looking at the energetic costs of the nuclear fuel cycle just from mining the ore through reactor construction to dismantling of the reactor, *without* even assessing the energy costs of storage and transportation of radioactive waste, the total energy debt comes to approximately 240 petajoules (24 million billion joules). The construction and implementation processes involved in a gas-fired plant require only one-tenth that amount—24 petajoules—to produce the same amount of electricity.[26]

Even utilizing the richest ores available, a nuclear power plant must operate at ten full-load operating years before it has paid off its energy debts. And, as noted above, there is only a finite supply of uranium ore containing reasonable concentrations of uranium 235. When this concentration falls below 0.01%, the costs of energy production from nuclear power can no longer cover the costs of extraction of uranium from the earth, at which time, the nuclear fuel cycle will deliver no net energy; below a certain uranium content, nuclear power produces less energy than is needed to build, fuel, and operate the reactor and to repair the environmental damage.[27]

Setting aside the energetic costs of the whole fuel cycle, and looking just at the Nuclear Industry's claim that what transpires in the nuclear plants is "clean and green," the following conditions would have to be met for nuclear power actually to make the substantial contribution to reducing greenhouse gas emissions that the industry claims is possible (this analysis assumes 2% or more growth in global electricity demand):

- All present-day nuclear power plants—441—would have to be replaced by new ones.

- Half the electricity growth would have to be provided by nuclear power.
- Half of all the world's coal fired plants would have to be replaced by nuclear power plants.[28]

This would mean the construction over the next fifty years of some 2,000 to 3,000 nuclear reactors of 1,000 megawatt size—one per week for fifty years! Considering the eight to ten years it takes to construct a new reactor and the finite supply of uranium fuel, such an enterprise is simply not viable.[29]

As van Leeuwen and Smith write, "the total known reserves of uranium . . . [are] so small one must ask oneself why it is that nuclear power was ever considered as holding promise of very large amounts of energy." They cite several possible reasons for this anomalous situation:

- The nuclear industry originally postulated that fast-neutron "breeder" reactors would be developed, which would create fuel as well as use it, in a self-sustaining "closed cycle." These reactors have yet to be realized.
- The industry did no calculations and had no conception of the huge energy costs associated with nuclear power.
- It was not understood for many years exactly how dangerous radioactive waste was and that long term disposal would be so intractable.
- It was not understood that uranium ores of less than 0.01% concentration could never deliver any net energy.
- All environmental damage induced by nuclear power was assumed to be left for future generations to rectify.[30]

With the knowledge about these topics that is now available, however, clearly the nuclear industry is running a public relations scam of massive proportions.

Disagreements exist about availability of uranium for the intended "nuclear renaissance." What differentiates the analysis in this chapter by Storm van Leeuwen and Smith is that no association or study group including the World Nuclear Association, has previously analyzed the uranium ore grade–energy relationship.

While highly-enriched military uranium is currently being mixed with low-enriched uranium for nuclear reactors in the United States, this amounts to only six years of the present annual natural uranium demand. There are no indications of new large rich deposits of uranium ore, and the currently known recoverable resources would supply 2,500 "renaissance" reactors for only eight years.

Some argue that reprocessing plutonium for reactor fuel will take care of deficient uranium supplies. Reprocessing is dangerous, extremely costly, and contributes to weapons proliferation.[31]

2

Paying for Nuclear Energy

THE TRUE ECONOMIC COSTS OF NUCLEAR ENERGY

The nuclear industry myth says that nuclear power costs only 1.7 cents per kilowatt hour, whereas coal costs 2 cents and gas-fired power costs 5.7 cents. But these figures apply only to nuclear energy generated from existing nuclear reactors. They represent a classic omission of capital costs from a pricing equation. Electricity from old nuclear reactors is relatively cheap because all the initial costs of construction, regulatory delays, and the like have been long forgotten.

A report from the New Economics Foundation titled "Mirage and Oasis: Energy Choices in an Age of Global Warming" concluded that the cost of nuclear power has been underestimated by almost a factor of three.[1] Indeed, the nuclear industry baldly provides false estimates of the financial cost of nuclear electricity, failing to account for the total nuclear fuel cycle, basing their numbers on incomplete data. Were the true costs to be factored in, the unit cost of nuclear power would rise dramatically.

An article in the *New Scientist* contradicts the nuclear industry assessment, concluding that once realistic construction and running costs are considered, the price of nuclear electricity rises from an estimated 3 pence per kilowatt hour in Britain (about 5 cents in the United States) to 8.3 pence (about 14 cents). This is because the price of construction of a new reactor is exorbitantly high, whereas it doesn't cost much to keep an old reactor running. This estimate still does not include the cost of managing pollution, accidents, insuring nuclear power stations, or protecting them from terrorists.[2]

As for creating more capacity by building new nuclear power plants, at the moment low interest rates may increase the appeal of big capital projects, but a rise in rates could put new construction on hold indefinitely.[3]

Using the current inflated price of natural gas as an argument for the economy of nuclear power is also fallacious. Ed Cummins from Westinghouse—a company that makes nuclear reactors—says that if natural gas goes back to its traditional price of $3.5 per MBtu from its current possibly anomalous price of $6, nuclear plants will cease to be competitive even if one accepts the industry's artificial estimates.[4] (Natural gas is so expensive at the moment because Hurricane Katrina severely disrupted supplies in the Gulf, the United States suffers from inadequate facilities to import large supplies of natural gas, and the overall global demand is increasing.)

Aside from natural gas comparisons, not all studies agree with the industry estimates. According to a 2003 study from the Massachusetts Institute of Technology titled "The Future of Nuclear Power," the "levelized" cost of electricity generated by a new nuclear power plant is about 60% higher than the cost of electricity from a coal-fired or combined cycle gas turbine plant, assuming moderate gas prices. (The levelized cost equals the total cost of electrical energy produced by a power plant divided by the total cost of construction and maintenance.) Even assuming the absence or underdevelopment of an existing grid for alternative electricity production, energy efficient technologies and renewable energy sources are now emerging as formidable financial competitors to nuclear power.[5]

SOCIALIZED ELECTRICITY

Perhaps the reason the leading nuclear power producers do not seem fazed by the high tab for nuclear power is that they do not expect to have to pick it up. Despite the fact that most developed countries are rhetorically wedded to principles of the "free market," their governments remain inexplicably enthusiastic about a form of energy that cannot be sustained without huge government subsidies and handouts. This socialization of nuclear electricity within a capitalist society has never been called into question, nor has it been critically scrutinized by the general public and their elected representatives. Yet dozens of the same CEOs calling for welfare reform with respect to individuals benefit from policies that amount to corporate welfare for the nuclear power industry.

One reason that nuclear power is so heavily subsidized relates to its history. Nuclear power was and still is an offshoot of the nuclear weapons industry and a beneficiary of the intensive government-funded research and development committed to nuclear weapons development during the Cold War. Although the nuclear power industry suffered an almost mortal economic blow after Three Mile Island and Chernobyl melted down, it is currently rising Phoenix-like, from the ashes of its own making. But nuclear energy's renaissance is based on a string of lies that begins by casting nuclear energy as the answer to global warming and goes on to position it as the economically responsible alternative to those "expensive renewable energy sources."

Chief among the untruths is information disseminated about the actual financial cost of nuclear energy and who will be footing the bill. Overt fabrications are propagated from the highest levels. On June 22, 2005, President Bush told workers at the Calvert Cliffs nuclear power plant, "One of the last things that we need to do to this economy is to take money out of your pocket and fuel government."[6] But, as we will see, the nuclear power industry's revival is utterly dependent upon taking money out of the taxpayer's pocket.

Before we examine the true economic realities of nuclear power it must be clearly stated to those investing millions of dollars in this technology that they will lose all, should there be a catastrophic nuclear meltdown in the United States or any other part of the world. Such an event would signal the end of nuclear power forever. A very experienced nuclear engineer, David Lochbaum, who works for the Union of Concerned Scientists is deeply concerned about the current lack of safety standards in U.S. reactors and is convinced there will be a nuclear catastrophe within the near future. He said to me, "It's not if but when." It seems, therefore, that it is a very risky business indeed to invest in nuclear power no matter what the industry or the government is currently saying.

Nuclear power has been and still is dependent upon government subsidies at every level. The U.S. government spent a gargantuan $111.5 billion on energy research and development between 1948 and 1998, allocating 60% or $70 billion of this to the nuclear industry alone.[7] Over the same fifty years, $26 billion was allocated to oil, coal, and natural gas; $12 billion went to renewable energy sources such as wind, hydro, geothermal, and solar power; and only $8 billion went to energy efficiency technologies.[8] In other countries, the Organisation for Economic Co-operation and Development (OECD) governments spent $318 billion by the year 1992 specifically on nuclear energy R&D.[9]

With this level of government support, it is no wonder that the nuclear power industry was wildly optimistic about its future by 1972, the year in which the Atomic Energy Commission predicted that the United States would have 1,000 nuclear power plants by the year 2000, as well as reprocessing plants to recycle spent fuel and breeder reactors that would produce as much nuclear fuel as they consumed. Dixie Lee Ray, then-head of the AEC arrogantly claimed that the disposal of spent nuclear fuel would be "the greatest nonproblem in history" and would be accomplished by 1985.[10]

In fact, the year 2000 saw completion of only 103 reactors, no operating breeders, no operating reprocessing facilities, and no high-level waste disposal sites whatsoever.

On the private financial side, the enormous expense of simply constructing a nuclear reactor—double the capital cost of a conventional coal plant—means that investors remain less than enthusiastic. The government has offered various incentives to try to engage private investment in the nuclear power industry. But even so, the rating agency Standard and Poor's recently concluded that "the industry's legacy of cost growth, technological problems, cumbersome political and regulatory oversight, and the newer risks brought about by competition and terrorism may keep credit risk too high for even federal legislation that provides loan guarantees to overcome."[11]

Caren Byrd, executive director of the global power and utilities group at Morgan Stanley, is cautiously optimistic, saying that for the first time in many years Wall Street believes that new nuclear reactors could become part of the nation's long-term energy future. She points out, however, that this forecast is largely dependent upon government support. She said that the Shoreham plant in New York was closed down after construction in 1985 because of enormous public opposition and that dozens of plants were cancelled in the 1980s, while others were plagued by long delays. "Tens of billions went down the drain at that time," she said. "We can't take that risk, and the investment community has long memories."[12]

In truth, the U.S. nuclear program in the past has been marred by construction cost overruns, delays, cancellations, premature plant closings, poor operational performance, and an inability to find a permanent storage site for its long-lived toxic radioactive waste. In one of the biggest cost overruns, the Seabrook reactor in

New Hampshire, which attracted huge grassroots opposition, was projected in 1976 to cost $850 million and to be completed in six years. It actually cost $7 billion, and fourteen years passed until its completion in 1990.[13] As one journalist sagely noted, when the nuclear industry declares success based solely upon the lack of greenhouse emissions, this attitude requires a very selective and pathological memory.[14]

According to Oxera, a British-based consultancy, even with an explicit tax on carbon-based power generation, new nuclear power plants cannot be economical without government subsidies. This estimate is supported by Britain's Royal Institute of International Affairs.[15] Optimistic rhetoric aside, no one has yet committed to building a new reactor, and a premature announcement could well rattle investors and depress a utility's stock, according to industry experts.[16]

DEREGULATION

If the economics of nuclear power look so dismal, why is there suddenly a huge push by the nuclear industry for new reactor construction? A little noticed but extremely important factor in the excitement for new nuclear power plants is a section hidden in the energy bill passed by the U.S. Senate in July 2005—a section repealing the Public Utilities Holding Company Act (PUHCA). Now, huge private power monopolies that were formerly excluded from owning nuclear plants can take over the small nuclear energy companies. This means that public utilities will disappear as they are bought by huge private enterprises, and the private companies will then have a monopoly that will make them so powerful they cannot be regulated. The rate payers and the public will be severely handicapped under this new system, just as they were before Roosevelt, under the New Deal, imposed strict federal regulations upon similar energy monopolies.

Signed into law in 1935, and a cornerstone of New Deal financial reform, PUHCA has legally and effectively controlled utility finances and operations for the better part of seventy years. PUHCA imposes strict state and federal government regulations, while simultaneously restricting ownership of utilities to those public or private entities that specifically produce power. This method of keeping speculators at bay was deemed necessary because, during the 1920s, utilities had become cash cows for energy moguls who set up complex holding companies. These companies were designed to milk income from the ever-present and ultra-reliable ratepayer to feed the moguls' speculative investments. The stock market crash of 1929 destroyed the holding companies, which not only devastated the rate payers, but also destroyed millions of small investors who had been sold the myth of a sure investment in utility reliability. The PUHCA laws designed to outlaw these structures were the biggest battle of FDR's first term.[17]

Lynn Hargis, who was employed for ten years at the Federal Energy Regulatory Commission, is extremely worried about the demise of PUHCA. She says that "it is clearly impossible for a state (or even a federal) utility commission, with its limited staff, to review, much less understand or control the books and records of a huge conglomerate." Once PUHCA is gone, she predicts "there will be a white hot fury of buying and selling utilities and utility assets—it will be a revival of the 1920s, when three huge companies owned half of all utilities."[18]

Who will be the first to benefit this time around? Halliburton, the famed oil company previously run by Vice President Dick Cheney, is perfectly positioned to take advantage of this deregulation and reap the benefits of formerly off-limits markets and captive rate payers. According to Hargis, "The top five oil companies now control 50% of U.S. oil production. If they also controlled public utilities, they would be too powerful for any government to regulate." Furthermore, she adds, the impact upon renewable energy could be devastating: "If GE owns your utility, nothing will be able to stop them from shoving a nuclear power plant down your throat. This will kill renewables. Not only is it going to be horrible for the whole country, but nobody is even talking about it."[19]

The top ten nuclear firms now own 61% of the nuclear sector, with Exelon owning the largest share of 15%.[20] In Europe, the energy market will be fully deregulated by 2007, and the U.S. energy market is in the process of deregulation, a process that will induce serious changes in the financial relationship between government and electricity producers.[21]

The repeal of PUHCA has already led to the rise of two huge consortia, and the U.S. government seems intent on pandering to them, giving them access to unheard of subsidies and changing the permit process in ways that favor the nuclear industry over consumer and citizen interests and well-being.

One of them, the Nustart consortium, is composed of Exelon, Entergy, Constellation Energy Group, Duke Energy Corp., EDF International North America—the North American division of Paris-based Electricite de France—Florida Power & Light Co., Progress Energy Inc., Southern Co., Tennessee Valley Authority, GE, and Westinghouse Electric Co.[22] Nustart has arranged for profits and liabilities to be dispersed among the companies, and the consortium hopes to receive site and design permits to build two new nuclear power plants by 2011 and to complete them by 2014. Half the estimated costs of the permit ($520 million) will be provided by the Department of Energy,[23] a socialized arrangement that suits the nuclear consortium.

Another huge consortium, Dominion Resources, is composed of an amalgamation of energy companies including Fairfield, Connecticut-based GE, design consultants in the Bechtel Group of San Francisco, and the U.S. Energy Department. This group is moving swiftly to obtain site and design permits for a new nuclear reactor to be constructed next to the Lake Anna nuclear power plant in Charlottesville, North Carolina. The permits, which will be allocated before construction even begins, will cost $440 million, half of which will be covered by the Department of Energy. Despite these quite extraordinary subsidies, which do not correlate with free market ethics, the senior vice president of Dominion stated that the biggest obstacle to building a new nuclear power plant is the economic risk, and the potential that at the eleventh hour an unforeseen legal challenge or regulatory hurdle could arise that would render the plant inoperable.[24]

In addition to the subsidies themselves, the government assists these consortia in myriad additional ways. For example, even though the DOE so generously underwrites these nuclear companies, there is no legal obligation on the part of the consortia ever to build a reactor. Furthermore, the permits obtained will be good for twenty years, they can be renewed for another twenty years, and they can be either banked or sold.[25]

The Bush administration has also surreptitiously modified the process of procuring permits for reactor construction to favor the industry by effectively excluding public intervention or participation as follows:

- Power companies can now apply for "early site permits," which offer a permit good for twenty years (renewable for another twenty years) as described above, without any public approval or input. Before these changes were made, the public had input at each stage of reactor licensing.
- Companies can now plug an approved plant design into a site with an early permit without further review or oversight.
- Companies can obtain a combined construction permit and operating license, issued at the time of construction,

thereby avoiding any public hearing in advance of plant operation.

Draconian in their reach, these measures effectively rule out any public opposition and intervention against new nuclear power plants. How can local communities effectively battle a huge conglomerate that has already obtained full approval processes for designing, building, and operating nuclear power plants issued twenty years in advance, when all forums for public discussion and input have been removed?

Since the Reagan years, the nuclear industry has maintained a near veto power over appointments to the Nuclear Regulatory Commission. Consequently, the commission's agenda converges ever more closely with the nuclear industry's. The NRC, ostensibly an independent body, now virtually represents the nuclear industry, which explains the above changes to the licensing laws.[26] Thus, the public has been effectively eliminated from the nuclear industry's decision-making process.

Apart from these egregious changes in the licensing process, Peter Bradford, former Nuclear Regulatory commissioner, outlines a further list of taxpayer-funded favors the government has offered the nuclear power industry:

- DOE will share half of all expenses in obtaining site permits and reactor licences
- Proposed construction financing guarantees
- Proposed tax deductions of 20% on the cost of building new reactors
- A proposed tax credit of 1.8 cents per kilowatt hour for all nuclear electricity generated—more than the industry says each kilowatt hour costs because as this chapter indicates, their data is unrealistic, and this bill is extraordinarily generous to the industry

• $1.8 million earmarked to assist the construction of "advanced" reactor designs

To help smooth the way, the NRC licensing process will not permit the *need* for electricity from the new plants to be challenged and will also exclude all challenges based on the uncertainty of the waste situation.[27]

BRAND-NEW SUBSIDIES

Let us now examine the lastest round of financial handouts for nuclear power in the context of standard justifications for subsidies. (These subsidies are embedded in the newly passed 2005 U.S. energy bill, an extraordinarily important, 1,700-page document, which few, if any, members of the House or Senate read thoroughly before they voted on it.)

Traditionally, energy subsidies have been used to reach economies of scale in the early years of a particular technology's development. But, according to the International Energy Agency, the technology associated with nuclear power is now "proven and mature."[28] On these grounds, subsidies for nuclear energy are anachronistic.

Standard economic theory states that subsidies can be justified when they lead to an overall increase in social welfare. But the environmental and health risks associated with radioactive waste, accidents, and risk of meltdown, nuclear proliferation, and the threat of terrorism *decrease* the overall contribution to social welfare provided by nuclear power.

The United Nations Environment Protection (UNEP) organization specifically dictates that the removal of subsidies that are economically costly and harmful to the environment and to people represents a win–win policy. It is hard to imagine a more poignant case in point than the nuclear power industry.[29]

Yet, the 2005 U.S. energy bill provides cradle-to-grave subsidies for nuclear power. The industry will gain $13 billion in subsidies and tax breaks, including:

- $5.7 billion in production tax credits;
- $4.4 billion in various subsidies—a conservative estimate that includes research and development, tax breaks, loan guarantees, and risk insurance;
- $1.25 billion from 2006 to 2021 and "such sums as necessary" from 2016 to 2021 for a nuclear power plant in Ohio to generate hydrogen for automobiles;[30]
- $435 million over three years for nuclear energy research and development, including the Department of Energy's Nuclear Power 2010 program to build new nuclear power plants and its Generation IV program to develop new reactor designs. Half the cost for new reactor applications, which are necessary to obtain governmental permission to build a new nuclear power plant—estimated to be $87 million for each reactor—will be covered by taxpayers.

The bill also includes unlimited taxpayer-backed loan guarantees, in a situation where the risk of a loan has been characterized by the Congressional Budget Office as "well above 50%." And it reauthorizes the Price-Anderson Act, with its guarantee that taxpayers, not the industry, will pay 98% of up to $600 billion governmental insurance in the event of a worst case nuclear meltdown. It is impossible to purchase personal or property insurance to cover a nuclear power plant accident. The industry could not function if it had to cover its own insurance.

These are the "direct" or "on-budget" subsidies guaranteed by the government that appear on the national balance sheets as government expenditure. Other large "on-budget" subsidies that do not apply to any other form of electricity generation, but must of necessity be provided to support the extraordinarily dangerous nuclear power industry, include the complex infrastructure necessary to transport nuclear waste; military protection against terrorism aimed at nuclear reactors and their cooling pools post 9/11; and naval escorts, which are essential for ships that transport nuclear waste.[31]

The nuclear industry is also the beneficiary of many other government subsidies that are hidden or "off-budget" items not immediately apparent to the taxpayers who pay for them.

These "off-budget" subsidies are numerous and include tax exemptions, credits, deferrals, rebates, preferential tax treatment, market access restrictions, regulatory support mechanisms, preferential planning consent, and access to natural resources.[32]

Several of the more offensive "off budget" subsidies include:

- Preferential tax treatment.
  In 1989, the British government withdrew nuclear power from its electricity sector privatization program because it was seen to be too great an investment risk. But a year later, a consumer subsidy equivalent to 50% of the government-owned Nuclear Electric's income in 1990 had to be introduced to assist the ailing nuclear sector because this nuclear utility could only recover half its costs from the sales of electricity to the new electricity market.[33]
- Stranded costs recovery.
  Because the cost of coal- and gas-fired electricity has been low (electricity is rarely generated from oil) and the price of nuclear-generated electricity has risen over time, many of the currently operating nuclear power plants are often unable to generate the rate of return predicted when the reactor was built. The difference between the predicted book value and the market value of such facilities is called a stranded cost. The nuclear industry argues that because the plants were built in good faith at a time when current market conditions did not exist, they should therefore be entitled to the income anticipated when the reactors were built.

The Federal Energy Regulatory Committee in the United States has accepted the general principle of stranded costs recovery as the nuclear industry moves to a competitive and deregulated market. Stranded costs by 1996 were estimated to be between $24 billion and $56 billion, and in 1999, the stranded costs in eleven states alone were estimated to be $110 billion!

In the United Kingdom before privatization, all nuclear energy assets were owned by the government. After privatization, the burden of stranded costs was shared between taxpayers and consumers—who are really one and the same thing. The benefits of stranded costs to the Spanish nuclear industry are estimated to be about 3 billion euros.[34]

• Liability Limitation

In most business situations individual risks are covered by an insurance policy and the business pays the premiums based upon the qualitative assessment of the risk involved. But in the United States, the Price-Anderson Act limits the nuclear industry's liability in the event of a catastrophic accident to $9.1 billion, which is less than 2% of the $600 billion guaranteed by the Congress. In any case, $600 billion is considered to be a gross underestimate; the actual dollar amount would depend on the area rendered uninhabitable, the extent of property damage, and the number of people killed and injured.

In the EU, the maximum economic liability set by the Paris Convention on Third World Liability and the Brussels Convention Supplementary to the Paris Convention was revised upwards in 2004 to 70 million euros,[35] an amount grossly inadequate to cover the cost of a major meltdown. In France, if Electricite de France had to insure for the full cost of a meltdown, the price of nuclear electricity would increase by about 300%. Hence, as opposed to conventional wisdom, the price of French nuclear electricity is artificially low.

- Decommissioning Funds
  Very few reactors have ever been decommissioned, but many plants are now reaching the end of their operating lifetimes and will need to be closed. As we have documented, this is an extremely expensive process. When the Yankee Rowe nuclear reactor in Massachusetts was decommissioned, the cost was projected to be $120 million; the actual cost was $450 million. (Reactors have become far more expensive; the Seabrook reactor in New Hampshire, completed in 1976, cost $7 billion to build, compared with its 1961 construction costs of only $39 million!)

  Because decommissioning has rarely been performed, there is a paucity of data to make projections. Timescales for decommissioning are very long, and added costs include problems with security, health, and the environment, possible damages to the local and national economies, and major financial risks should there be an accident with intensely radioactive parts.
- External Costs
  The environmental and health costs associated with the nuclear industry, plus the risks of nuclear weapons proliferation and terrorism are automatically transferred from the private to the public domain. The nuclear industry is not expected to accept any fiscal responsibility for these "externalized" risks.[36]

Other subsidies deemed "shadow R&D" include monies spent by public institutions such as universities and nuclear physics institutions on extensive nuclear research that is funded by general research grants. Still others are embedded in the U.S. Energy supply appropriation bill, which tends to evade public scrutiny. The Office of Nuclear Energy, Science and Technology (NE), for example, operates ten programs, all of which are directly associated with the development and use of nuclear power.

These programs have been incorporated in the agenda of a huge nuclear infrastructure established during the Cold War specifically for nuclear weapons research, development, testing, and production, of which research on nuclear power was always an integral part. The sprawling nuclear weapons laboratory complex, which litters the nation from east to west, north to south, includes the Los Alamos National Laboratory in New Mexico, the Lawrence Livermore National Laboratory in California, Sandia National Laboratories in New Mexico, the Oak Ridge National Laboratory in Tennessee, and Pacific Northwest National Laboratories in Washington State.[37] Many of these facilities have recently become deeply involved in the nuclear power renaissance.

The intricacy, complexity, and sheer vastness of this nuclear network is enough to take one's breath away. It is so well funded, so entrenched, so organized, so out-of-sight, that it is difficult to imagine how it could possibly be dismantled for the betterment of the human race.

The following paragraphs, numbered 1–10, have been lifted almost verbatim from the 2005 U.S. congressional budget on energy.[38] They describe what amounts to an enormous, free R&D arm for the nuclear energy industry, based at universities and labs across America. Needless to say, none of the costs involved in maintaining any of these programs appears on any balance sheet or in any financial calculations provided by the nuclear industry in making the case for nuclear power. (If the following paragraphs seem complex and difficult to understand, remember that this is the language used by the Department of Energy. Much of this technological gobbledegook will be translated and explained in later chapters on the operation and debilities of old nuclear reactors and the chapter describing the new generation of nuclear reactors that the nuclear industry is enthusiastically planning to foist upon an unsuspecting and energy-hungry public.)

1. *The University Reactor Infrastructure and Education Assistance Program* "supports the upgrade of university research and training reactors, provides graduate fellowships and undergraduate scholarships to outstanding students, uses innovative programs to bring nuclear technology to small, minority serving institutions, and provides nuclear engineering research grants to university faculty. The program helps to maintain domestic capabilities to conduct research and critical infrastructure necessary to attract, educate, and train the next generation of scientists and engineers with expertise in nuclear technologies. DOE also provides the supply of fresh fuel to university research reactors and supports reactor equipment upgrades at universities." [Cost: $21,000,000]

2. *The Nuclear Energy Plant Optimization* is organized by the Argonne National Laboratory (ANL) and will assess the effectiveness of non-destructive examination techniques for the detection and characterization of service-induced cracks in steam generator tubes and will provide ongoing support of signal validation technologies and quantification of benefits of online monitoring. [No data on cost available]

3. *The Nuclear Energy Research Initiative* is supervised by ANL and consists of nine projects in the areas of reactor systems, fundamental chemistry, material science for Generation IV systems, integrated nuclear and hydrogen production, and advanced nuclear fuels/fuel cycles. [No data on cost available]

4. *Nuclear Energy Technologies* is run by ANL. This particular program consists of a macroeconomic assessment relating to the development of new nuclear power plants. [Cost: $10,246,000]

5. *The Generation IV Nuclear Energy Systems Initiative.* Argonne National Laboratories (ANL) and the Idaho National Laboratory (INL) will coordinate the preparation of the Generation IV Technology Roadmap. ANL will coordinate with France and with Korea in advanced conventional methods, gas cooled reactor

technology, and advanced fuels and materials with the International Nuclear Energy Research Initiative (I-NERI). ANL will also work on the melt/concrete interaction, meaning they will study what happens to the concrete container in the event of a meltdown. [Cost: $30,546,000]

6. *The Nuclear Hydrogen Initiative,* again run by ANL, will conduct laboratory analyses of nuclear propelled thermochemical hydrogen production methods, specifically the calcium-bromine cycle. [Cost: $9,000,000]

7. *The Advanced Fuel Cycle Initiative.* Yet again, ANL will perform reactor physics calculations, including spent fuel throughput calculations for existing commercial light water reactors and Generation IV thermal and fast reactor concepts. [Cost: $46,254,000]

8. *Radiological Facilities Management* is essentially the linear accelerator at Brookhaven National Labs located in the middle of Long Island, which injects 200 million-electron-volt protons into a 33 giga-electron-volt Alternating Gradient Synchroton. Some medical isotopes are produced at this facility. [Cost: $69,110,000]

9. *Idaho Facilities Management and Program Assessment.* Idaho National Labs have been heavily involved in nuclear research for many years. Now they are managing, among many other nuclear projects, the Advanced Test Reactor (ATR), one of the world's largest and most advanced test reactors, which provides vital irradiation testing for reactor fuels and core components, primarily for the U.S. Navy's Nuclear Propulsion Program. It also produces isotopes for medicine and industrial purposes. Other facilities at INL include the ATR Critical Facility reactor; the ATR Hot Cells, which are used to manipulate plutonium; the Office of Science's Safety and Tritium Applied Research Facility (STAR), which performs fusion fuel research, and the INL Applied Engineering and Development Laboratory. [Cost: $87,164,000]

10. *Idaho Sitewide Safeguards Band Security.* This is essentially a security operation that provides protection for nuclear materials,

classified information, government property, cyber security, and personnel security. [Cost: $58,103,000]

The subsidies received thus far for renewable energy sources have been dwarfed by those provided to the fossil and nuclear sectors. In the United States alone, for the first fifteen years of its development, the nuclear sector received thirty times as much financial support—$15.3 per kilowatt hour (kWh) compared with a measly $0.46 per kWh for wind energy development.[39] For the same amount of investment, wind power creates five times as many jobs and generates 2.3 times as much electricity as nuclear power.[40] (Other renewable energy sources will be examined in a later chapter.)

To summarize, the actual costs of nuclear energy are consistently misstated and incomplete. Nuclear power is also heavily subsidized by taxpayers (through programs that benefit the industry, but are excluded from their cost estimates). Developed countries ostensibly wedded to the principles of economic rationalism and the "free market," are inexplicably enthusiastic about nuclear power, which cannot be sustained without huge government subsidies and handouts from its very inception. This socialization of electricity within a capitalist society has never been called into question, nor has it been critically scrutinized by the general public and their elected representatives.

3

Nuclear Power, Radiation, and Disease

Few, if any, estimates of the costs of nuclear energy take into account the health costs to the human race. Even when nuclear power plants are operating normally, these costs are not insignificant. Miners, workers, and residents in the vicinity of the mining and milling functions, and workers involved in the enrichment processes necessary to create nuclear fuel are at risk for exposure to unhealthy amounts of radiation and have increased incidences of cancer and related diseases as a result. Routine and accidental radioactive releases at nuclear power plants as well as the inevitable leakage of radioactive waste will contaminate water and food chains and expose humans and animals now and for generations to come. Accidents such as Three Mile Island and Chernobyl condemn thousands if not millions to pay the cost of nuclear power with their own health. Understanding the nature of radiation is critical to understanding the health impacts of nuclear energy.

RADIATION AND EVOLUTION

Billions of years ago the earth was relatively radioactive and hostile to life, as radiation emanated both from the terrestrial plane—rocks and soil—and from powerful solar radiation in space. The solar effect was much more intense at that time, because the ozone layer that filters out the carcinogenic ultraviolet radiation from the sun was almost non-existent. Over billions of years, as plants developed and evolved, they generated oxygen (O_2) from photosynthesis, which rose up through the lower layers of atmosphere into the stratosphere, where it was converted to ozone (O_3) by ultraviolet light. Gradually, as the ozone layer accumulated in the upper atmosphere, the intensity of the solar radiation diminished, as did the terrestrial radiation, and the earth "cooled down."

Life began as primitive single-celled organisms, which, over billions of years evolved into many and varied multicellular organisms. Human beings appeared relatively recently—only 3 million years ago. Background radiation was one of the main instigators of evolution, as it induced mutations in the reproductive DNA molecules or genes of plants and animals. (A mutation is a biochemical change in the double helix DNA molecule.) The vast majority of mutations were "deleterious," causing death and disease in the offspring, but some were "advantageous," allowing the new organisms to flourish in a hostile and difficult environment. Fish developed lungs and climbed out of the water to become land-dwelling amphibians, dinosaur-like creatures developed wings, and became the earliest form of birds, and humans evolved as our predecessors stood on their hind limbs, grew the opposing thumb, and developed a huge cerebral neocortex—changes that eventually allowed us to dominate and control the natural environment. As the earth cooled down and background radiation decreased, genetic mutations decreased in frequency, and species adjusted to this change.

Radiation, which has been fundamental to the evolution of planetary life, is largely responsible for development of the most extraordinary and wonderful variety of living species over a time frame of billions of years. But humans seem determined to alter this stable balance bequeathed to us by nature. With only the barest comprehension of evolution or the delicate process of genetics, we create massive quantities of radioactive elements to power our "lifestyle" because we are attached to ever-increasing levels of technological progress, prosperity, luxury, and ease of living.

When humans succeeded in splitting atoms, they also embarked upon a process that would inevitably increase the levels and diversity of background radiation on the earth. The process of fissioning uranium in nuclear reactors creates more than 200 new, man-made radioactive elements. Some "live" for only seconds; some remain radioactive for millions of years.

Once created, these diabolical elements will inevitably find their way into the environment and will eventually enter the reproductive organs of plants, animals, and humans, where they will mutate the genes in reproductive cells to cause disease and death in the immediate generation or pass a hidden genetic disease to distant offspring down the time track. This is because, as explained above, most mutations cause disease, whereas advantageous mutations are infrequent and require millions of years to express themselves.

RADIATION AND HUMAN REPRODUCTION

As many of us have learned in biology courses, genes are composed of DNA molecules, which are the very building blocks of life, responsible for every inherited characteristic in all species—plants, animals, and humans. Every gene in an egg or sperm is precious and unique. When the egg and sperm are created, the number of genes in each reproductive cell is halved, so that when conception occurs, the new individual has a full complement of genes. Most characteristics are governed by a pair of genes, one inherited from the mother, and one from the father. Each of these genes can be either dominant or recessive.

The characteristics of dominant and recessive genes are well demonstrated by eye coloring: Brown-eyed genes are dominant and blue-eyed genes are recessive. Blue eyes can manifest only if a person inherits of *pair* of blue-eyed genes—this individual is homozygous for blue eyes. Brown-eyed people can be either homozygous with two brown-eyed genes or heterozygous with a brown- and a blue-eyed gene. If, for example, the mother has blue eyes and the father is homozygous for brown eyes, all offspring will be brown eyed because all the sperm will carry the brown-eyed gene. But if the father is heterozygous, the baby has a 50% chance of having brown eyes, because half the sperm will carry the blue-eyed gene and half the brown-eyed gene.

The same holds true for many inherited diseases. Cystic fibrosis (CF), the most common lethal genetic disease of childhood, is inherited as a recessive gene. One in twenty-five people of Caucasian descent carry this recessive gene. When two CF carriers mate, there is a one in four chance that two CF genes will unite to produce an offspring with CF, that is, one fetus will be homozygous for two normal genes, two will be heterozygous for a normal gene and one for CF, and one will be homozygous for two recessive CF genes.

All people carry recessive genes for disease, but not until a person mates with another carrier of one of their abnormal genes can a baby be born manifesting this disease. The majority of these abnormal genes have been caused by mutations in the distant past—usually by background radiation, although some new mutations do arise spontaneously.

The seminal work on radiation and genetics was performed in 1927 by Dr. H.J. Muller, who irradiated drosophila fruit flies. Because these flies reproduce very rapidly, Muller could observe the effects on hundreds of generations within a short space of time. For instance, a radiation-induced dominant mutation that caused a crooked wing would be passed down through many generations of fruit flies. Dr. Muller was awarded a Nobel Prize for his pioneering work. Other researchers have since verified Muller's findings, and the number of mutations has been shown to be in direct ratio to the cumulative amount of radiation received by the reproductive organs, be it a single large dose or many smaller doses.

Radiation induces mutations that are either dominant, recessive, or sex linked—carried on the female X chromosome or in cellular mitochondria, which determine some genetic characteristics. Many conditions such as diabetes, cystic fibrosis, muscular dystrophy, and certain forms of mental retardation are recessive diseases. Two typical sex-linked genetic diseases are color blindness and hemophilia. There is a total of 16,604 genetically inherited diseases now described in the literature.[1]

All human cells have forty-six chromosomes in their nucleus, and genes themselves are arranged in pairs along twenty-three pairs of chromosomes. Apart from *genetic* mutations, radiation can cause breaks in *chromosomes*, which can cause a baby to be born with Down's syndrome or some other serious mental or physical disorders. A normal fetus with fully functioning genes and chromosomes can also be damaged by external radiation exposure or if a radioactive element crosses the placenta and lodges in the fetus, killing a particular cell that would later form the septum of the heart, the right half of the brain, or the left arm, for example. This pathological process, which results in malformations of the heart, the brain, the limbs, or other organs of a fetus, is called teratogenesis. Similar deformities were observed in decades past when pregnant women took the drug thalidomide—which likewise killed important cells within the fetus—to alleviate their morning sickness.

RADIATION AND DISEASE

All non-reproductive or "somatic" body cells have regulatory genes that control the rate of cell division. If a regulatory gene is biochemically altered by radiation exposure, the cell will begin to incubate cancer, during a "latent period of carcinogenesis," lasting from two to sixty years. Then one day, instead of the cell dividing into two daughter cells in a regulated fashion, it will begin to divide in a random, uncontrolled fashion into millions and trillions of daughter cells, creating a cancer. Cancer cells tend to be very invasive. They break off from the main cancer mass, invading lymph vessels and blood vessels in a microscopic fashion, and travel to other organs (liver, bone, lung, brain, etc.) where they grow into secondary cancers or metastases. In many cases it is difficult if not impossible to stop this random growth of abnormal cells. Thus, a single mutation in a single gene can be fatal.

It is thought that 80% of cancers that we see are caused by environmental factors, whereas only 20% are inherited. Cancer has always plagued the human race; some ancient Egyptian mummies were riddled with cancers. It is generally accepted that many cancers in the past and in the present have been and are caused by background radiation. Because aging exposes people to increasing doses of radiation and carcinogenic chemicals, cancer is generally a disease of old age.

However, no dose of radiation is safe, and all radiation is cumulative. Each dose received adds to the risk of developing cancer or mutating genes in the reproductive cells. (The risk is small and the benefit great when a serious diagnosis must be made, but exposure to unnecessary X-rays or CAT scans must be avoided.) We are exposed to a background radiation dose of about 100 millirems per year from the earth and the sun. It has been estimated that if one hundred and twenty-five people receive 100 millirems per year for seventy years, one of them will develop cancer. But the Nuclear Regulatory Commission (NRC), which is responsible for the oversight of the nuclear power industry, has decided that it is acceptable for the public to receive an additional 100 millirems per year from man-made radiation created through the generation of nuclear energy, meaning two extra cancer patients will be created out of every hundred people annually, adding together the one cancer from background and one from "allowable" man-made radiation.[2]

The rules are even more lenient for nuclear workers, who are allowed doses of 5 rems per year (5,000 millirems). One in five nuclear workers are predicted to develop cancer if they received this "legally allowable" dose over fifty years of exposure.[3] These workers have to operate in areas that are very radioactive or "hot," exposing their reproductive organs to radiation. Because most nuclear workers are men, mutated genes in their sperm will be inherited by their offspring and passed on to future generations. The few women nuclear workers will be similarly affected as genes mutate in their eggs. The nuclear industry cannot function without these dangerous exposures, but one wonders if any nuclear workers are adequately informed about the biological dangers of working in the nuclear industry.

Furthermore, when the nuclear industry calculates "acceptable" radiation exposure for the public, they use a model of a standard, healthy 70 kilogram man. But the population is far from homogeneous. Old people, immuno-depressed patients, normal children, and some with specific, inherited diseases are many times more susceptible to the deleterious effects of radiation than normal adults. Overall, about forty-two people out of hundred are expected to develop cancer in their lifetimes from all causes. Children born to parents who have been exposed to radiation have a higher-than-normal risk of developing cancer or leukemia.[4] High levels of radiation are also known to cause heart disease and strokes.[5]

The incidence of cancer in adults is on the rise,[6] particularly cancers of the kidney, brain, and liver; non-Hodgkin's lymphoma; and testicular cancer. Children have experienced an elevated cancer incidence as well, particularly of brain cancers,[7] as we pollute the environment with carcinogenic chemicals and radioactive elements. Eighty thousand different chemicals are in common use, very few of which have been tested for carcinogenicity. Chemicals and radioactive elements tend to act synergistically in human and animal bodies—one will potentiate the carcinogenic effect of the other.

According to a National Academy of Sciences report,[8] man-made radiation in the United States accounts for 18% of human exposure. Other sources of radiation include exposure to naturally occurring radioactive radon gas, to radioactive rocks and minerals on the earth, and to ultraviolet radiation from the sun. Of the man–made category, medical X-rays and nuclear medicine (short-lived radioactive elements used in diagnostic examinations and for the treatment of some cancers) account for about 79%, whereas radioactive elements in consumer products such as tobacco, tap water, and nuclear power currently account for 5%.[9] But this is now. As the huge quantities of radioactive waste accumulating from nuclear power and from nuclear weapons production start leaking and contaminating drinking water and food chains in many parts of the world, so the percentage of radiation exposure from these sources will rise.

In summary, the 18% human exposure attributable to man-made radiation will increase, because radioactive waste remains potent for hundreds and thousands of years. So by turning on our lights today, we bequeath our descendants a radioactive legacy for tomorrow.

ROUTINE RADIATION FROM NUCLEAR POWER PLANTS

Before we consider radioactive elements that are released from the nuclear fuel cycle, we must first define what kind of radiation they emit and what sort of damage they may do to living cells. Each radioactive element or isotope is unique in its physical properties and has a specific half-life. For example, radioactive iodine 131 has a half-life of eight days, so that in eight days it loses half its radioactive energy, in another eight days it decays again to one quarter of the original radiation, ad infinitum. It is customary to multiply a half-life by roughly twenty to calculate the time that a particular isotope will retain its radiation. In the case of iodine 131, its radioactive life therefore is 160 days or twenty-three weeks.

Some isotopes made in a nuclear reactor have very short half-lives (less than a second) and some extremely long (millions of years). These isotopes also emit several types of radiation. Many emit gamma radiation, which is akin to X-rays. Gamma radiation goes straight through human bodies. It does not make a body radioactive, but as gamma rays pass through the body, they can mutate regulatory or reproductive genes.

Some of the new isotopes emit alpha radiation, which is a particle composed of two protons and two neutrons shot out from an unstable atomic nucleus. The nuclear industry has said that alpha radiation is not dangerous because it doesn't travel very far and can be stopped by a piece of paper. Likewise it does not penetrate the layers of dead cells in the human skin or epidermis to damage living cells. However, if it enters the body through the gastrointestinal tract or is inhaled into the lung, it comes into direct contact with living cells and, as such, is extremely mutagenic.

Other isotopes emit beta radiation, which is composed of an electron shot out from an unstable nucleus. Beta radiation travels farther than alpha because it is lighter. It too is very mutagenic and carcinogenic.

The radiation given off by isotopes is insidious and cryptogenic (hidden). Various radioactive elements become incorporated into specific organs of the body. For instance, if you inhale one-millionth of a gram of the alpha emitter plutonium, a very small volume of cells in the lung is irradiated because of the very short distance travelled by the alpha particle. Because alpha radiation is so deadly, most of the cells within the radiation field will be killed, but as radiation decreases with the square of the distance, cells on the periphery of the radiation field remain viable. Some of them almost certainly will suffer mutation of their regulatory genes, and cancer will later develop in one of these damaged cells.

There are many routes of exposure to man-made radiation from the nuclear industry. Relatively small but significant amounts of radiation are released on a daily basis into the air and water during the course of mining, milling, and enriching uranium for fuel to create the nuclear energy. Additionally, a nuclear power plant cannot operate without routinely releasing radioactivity into the air and water through the normal operation of nuclear reactors. Finally, and most frighteningly, accidental releases of even more radiation are commonplace in the nuclear industry.

Uranium Mining

Uranium mining began in Europe in the late part of the nineteenth century when Madam Curie was refining pitch blend from uranium ore and separating radium. Large-scale mining commenced sixty-five years ago specifically to provide fuel for nuclear weapons and continued unabated for many decades thereafter. Over one-half of all uranium deposits lie under Navajo and Pueblo tribal land, in the United States[10] and over the years, large numbers of Native Americans have been employed as below-ground and above-ground miners.

People who mine uranium below the ground are at great risk because they are exposed to a high concentration of radioactive gas called radon 220, which accumulates in the air of the mine. Radon is a daughter or decay product of uranium and is a highly carcinogenic alpha emitter, which, if inhaled, can decay in the lung and deposit in the air passages of the lung, irradiating cells that then become malignant. As a result, uranium miners have suffered from a very high incidence of lung cancer. One-fifth to one-half of the uranium miners in North America, many of whom were Native Americans, have died and are continuing to die of lung cancer.[11] Records reveal that uranium miners in other countries, including Germany, Namibia, and Russia, suffer a similar fate.[12]

Another lethal uranium daughter is radium 226, which is an alpha and gamma emitter with a half-life of 1,600 years. This radioactive element is notorious in the medical literature. In the early part of the twentieth century, women painted numbers on watch dials with radium enriched paint, so that the numbers glowed in the dark with radioactivity. To make the figures precise, they licked the tips of the paint brushes, thereby swallowing large amounts of radium. Because radium is a calcium analogue, it deposited in their bones. Many of these women subsequently died of osteogenic sarcoma, a highly malignant bone cancer affecting their facial bones, whereas others succumbed to leukemia, because white blood cells are manufactured in the bone marrow. Uranium miners are exposed to a similar risk because radium is an integral component of uranium dust in the mine. When they swallow the dust, radium is absorbed from the gut and deposits in their bones. Uranium itself also deposits in bone, and it too is carcinogenic.

Uranium ore also emits gamma radiation, which emanates from the ore face. So the miners are also exposed to a constant, whole-body radiation (like X-rays) emitted by other uranium daughters, which irradiates their bodies and continuously exposes their reproductive organs.

As the uranium ore is mined and the uranium is extracted, large quantities of radioactive dirt and soil are discarded and left lying in huge heaps adjacent to the mine, exposed to the air and the rain. This material is called tailings. Most tailings in North America are situated on indigenous tribal land of the Navajo nation and the Laguna Pueblo in New Mexico and on the Serpent River First Nation in Ontario, Canada. By 1980, the sovereign Navajo nation had forty-two uranium mines and seven mills located on or adjacent to reservation or trust land. Millions of tons of radioactive dirt constantly leak radon 220 into the air, exposing the indigenous populations who live nearby. As they inhale the radon, many of these people have developed or are developing lung cancer.[13]

Rain also leaches soluble radium 226 through the tailings piles into the underground water,[14] which is often the source of drinking water. When radium enters streams and rivers, it bio-concentrates tens to hundreds of times at each step in the food chain of the aquatic life and terrestrial plants. Because it is tasteless and odorless, people in these contaminated populations cannot tell whether they are drinking radioactive water, breathing radioactive air, or eating fish or food that will induce bone cancer or leukemia.

Hundreds of mines and tailings heaps lie exposed to the air and wind on Navajo land. Thousands of Navajos are still affected by uranium-induced cancers and will continue to be so for thousands of years unless remediation takes place.[15] In total, 265 million tons of uranium tailings pollute the American Southwest.[16] Neither the government nor the nuclear industry has ever attempted to clean up this massive radioactive pollution of tribal land. It is hard to imagine, however, similar piles of radioactive tailings lying adjacent to the well-heeled town of New Canaan, Connecticut, or near the Rockefeller estate in the Adirondacks.

Uranium Milling

The U.S. federal government covers the cost of milling uranium, the process by which the mined ore is crushed and chemically treated to convert the uranium metal into a compound called yellow cake. As in the mining process, the waste ore is discarded on the ground, primarily on Navajo tribal land in the American Southwest, where the government mills are situated. These mill tailings contain radium and a dangerous radioactive element called thorium—a uranium daughter and an alpha and gamma emitter with a half-life of 80,000 years. Over the last forty years, over 100 million tons of mill tailings have accumulated mainly in the Four Corners area (the intersection of Arizona, Colorado, New Mexico, and Utah) in the American Southwest.[17]

In the mid-1960s, local contractors at Grand Junction in Colorado discovered acres of discarded mill tailings, unguarded and untreated. Not knowing they were radioactive, the contractors used them for cheap landfill and in concrete mix. Schools, hospitals, private homes, roads, an airport, and a shopping mall were constructed using this material. In 1970, local pediatricians noticed an increased incidence of cleft lip, cleft palate, and other congenital anomalies among newborn babies born to parents who lived in these radioactive structures, which continually emitted gamma radiation and radon gas.[18]

The EPA allocated monies to the University of Colorado Medical Center to study the correlation between the birth defects and the radioactive dwellings. However, one year into the study, funds were abolished because federal authorities claimed that the government had to cut back on many programs for budgetary purposes.[19]

Uranium Enrichment

As described in chapter 1, the uranium 235 isotope is enriched from a low concentration of 0.7% to 3% for fuel in nuclear power plants. (If uranium 235 is enriched above a concentration of 50%, it can be used as nuclear weapons fuel.) Workers at all stages of the enrichment process are exposed to whole-body gamma radiation from by-products of uranium decay. But the most serious aspect of enrichment is the material that is discarded: uranium 238. This is called "depleted uranium" (DU) because it has been depleted of its uranium 235. But it is not depleted radioactively.

Depleted uranium is lying around in thousands of leaking, disintegrating barrels at the enrichment facilities in Paducah, Kentucky; Oak Ridge, Tennessee; and Portsmouth, Ohio. At Padacah alone, some 38,000 cylinders of DU await disposal. DU has contaminated the ground water, forcing the government to provide alternative drinking water for the local residents.[20]

The Pentagon, however, has found a nifty use for at least a small amount of this radioactive waste. Because uranium 238 is 1.7 times more dense than lead, it has been deemed the ideal antitank weapon. When shot out of a cannon, the solid uranium antitank shell cuts through the steel on the other fellow's tank like a hot knife through butter. But DU has several unfortunate properties. It is pyrophoric, which means that it bursts into flame upon impact, and when it burns, up to 80% disintegrates into finely powdered aerosol, which is distributed to the four winds. The mist is radioactive, and it has a half-life of 4.5 billion years.

Uranium is a heavy metal. It enters the body via inhalation into the lung or via ingestion into the GI tract. It is excreted by the kidney, where, if the dose is high enough, it can induce renal failure or kidney cancer. As a calcium analogue, it lodges in bones where, like plutonium, it causes bone cancer and leukemia. Last but not least, it is excreted in the semen, where it mutates genes in the sperm.

In the 1991 Gulf War invasion, the Pentagon used 360 tons of depleted uranium in the form of antitank shells in Iraq, Kuwait, and Saudi Arabia.[21] In the invasion that began in 2002, the United States and its allies have already deployed well over 127 tons, according to the Pentagon, which is loath publicly to announce the total amount of DU used. I suspect the actual quantities are significantly higher. Much of the DU is in cities such as Baghdad, where half the population of 5 million people are children who play in the burned-out tanks and on the sandy, dusty ground. Children are ten to twenty times more susceptible to the carcinogenic effects of radiation than adults. My pediatric colleagues in Basra, where this ordnance was used in 1991, report a sevenfold increase in childhood cancer and a sevenfold increase in gross congenital abnormalities.

In essence, the two Gulf wars have been nuclear wars because they have scattered nuclear material across the land, and people—particularly children—are condemned to die of malignancy and congenital disease essentially for eternity. Because of the extremely long half-life of uranium 238, the food, the air, and the water in the cradle of civilization have been forever contaminated.

It is important to note that other countries involved in uranium enrichment include Britain, China, Russia, Israel, Japan, Germany, Argentina, France, North Korea, Iran, Pakistan, Brazil, and India. Iran is not building nuclear weapons now, and the scale of North Korean and Israeli enrichment programs is not clear. Britain, China, Russia, France, and Pakistan have highly-enriched uranium weapons. Many of these other countries, if they so desired, could make nuclear weapons by enriching their uranium beyond 50%. America set the example years ago, and the world follows.[22]

Fuel Fabrication

The fabrication of nuclear fuel involves more human exposure to radioactive materials. After milling, the uranium fuel is made into cylindrical ceramic pellets the size of a cigarette filter and placed in hollow zirconium fuel rods, half-an-inch thick and twelve-to-fourteen feet long. Each rod contains at least 250 pellets. About 50,000 of these rods are then packed into the core of a thousand megawatt reactor within a cylindrical space, fourteen feet high and twenty feet in diameter. Fuel fabrication workers are once again exposed to gamma radiation emanating from the uranium, as well as to radon gas and uranium dust.[23]

Routine Releases from Operation of Nuclear Power Plants

One hundred tons of uranium are placed in the core of a 1,000 megawatt nuclear power plant and immersed in water. When uranium is packed tightly together and the moderating rods made of boron are slowly removed, the uranium reaches critical mass.

Neutrons ejected from the atoms hit other uranium atoms which then break apart, ejecting more neutrons. A by-product of this process is the creation of over 200 new radioactive elements that didn't exist until uranium was fissioned by man.

The resulting uranium fuel is a billion times more radioactive than its original radioactive inventory.[24] A regular 1,000 megawatt nuclear power plant contains an amount of long-lived radiation equivalent to that released by the explosion of 1,000 Hiroshima-sized bombs. This process inevitably entails the release of radioactive materials into the environment. Over time the uranium swells. Pinhole breaks appear in the zirconium cladding, and some faulty welds rupture in the zirconium fuel rods themselves, releasing radioactive isotopes or elements into the cooling water. In addition, radiation emitted through the wall of the fuel rods activates water molecules and creates radioactive elements in the water itself. For example, neutrons emitted from the fuel rods interact with water molecules to form tritium—a radioactive isotope of hydrogen. The primary coolant—water that cools the reactor core—thus becomes intensely radioactive.

This thermally hot primary coolant is piped through a steam generator to heat the secondary cooling system. This secondary water is converted to steam, which turns the generators to produce the electricity. The primary coolant is not supposed to mix with the secondary coolant, but it routinely does, allowing radiation to be released to the environment from this secondary system.

Radioactive gases that leak from fuel rods are also routinely released or "vented" into the atmosphere at every nuclear reactor. These gases are temporarily stored to allow the short-lived isotopes to decay and then released to the atmosphere through engineered holes in the reactor roof and from the steam generators. This process is called "venting." About 100 cubic feet of radioactive gases are also released hourly from the condensers at the reactor. Planned ventings increase in frequency when the reactor shuts down due to mechanical malfunctions. Accidental ventings are not infrequent.[25]

Planned "purges," when radioactive gases are actively flushed into the atmosphere by a fan, are officially permitted by the NRC so that utility operators can decrease the intensely radioactive environment into which maintenance workers must enter. Older reactors are allowed twenty-two purges per year during routine operation and two purges per year during cold shutdown.[26] (Cold shutdown occurs when the fission reaction is stopped at the reactor and 30 tons of very radioactive fuel is removed and replaced by new fuel).

Some of the more dangerous gases, such as iodine 131, are usually trapped by filters, but not always. After the radioactive iodine is filtered, noble gases are routinely released. The nuclear industry argues that noble gases are chemically inert and therefore not capable of reacting biochemically in the body, but they actually decay to daughter isotopes, which themselves are chemically very reactive.

Noble gases have names that bring to mind Superman—xenon, argon, krypton. There are many varieties of these elements, some of which are described below. Noble gases are high-energy gamma emitters, and they are readily absorbed from the lung and enter the blood stream. Although they are chemically non-reactive, they are very fat soluble, and they tend to locate in the abdominal fat pad and upper thighs, adjacent to the testicles and ovaries. There, they can induce significant mutations in the eggs and sperm of the people living adjacent to a reactor.[27]

There have never been any epidemiological studies performed on the effects of exposure to the noble gases xenon and krypton.[28] This is a grave deficit in the study of radiation biology, because these gases are so ubiquitous around nuclear reactors and are released with irresponsible impunity. Several of the more dangerous isotopes to which noble gases decay (all of which have different metabolic pathways in the body) include the following:[29]

- Xenon 137, with a half-life of 3.9 minutes, converts almost immediately to the notoriously dangerous cesium 137 with a half-life of thirty years.
- Krypton 90, half-life of 33 seconds, decays to rubidium 90, half-life of 2.9 minutes, then to the medically toxic strontium 90, half-life of twenty-eight years.
- Xenon 135 decays to cesium 135 with an incredibly long half-life of 3 million years.
- Large amounts of xenon 133 are released at operating reactors, and although it has a relatively short half-life of 5.3 days, it remains radioactive for 106 days.
- Krypton 85, which has a half-life of 10.4 years, is a powerful gamma emitter.
- Argon 39 has a 265-year half-life.

Other dangerous noble gases include xenon 141, 143, and 144, which decay to cerium 141, 143, and 144. According to the National Council on Radiation Protection (NCRP Report No. 60) these three cerium isotopes, which are beta emitters, are abundant products of nuclear fission reactions and have moderately long half-lives. They bio-concentrate in the food chain, and they irradiate the lung, liver, skeleton, and gastrointestinal tract, where they act as potent carcinogens.[30]

A very important and little-discussed isotope that is routinely emitted in large quantities into the air and waste water from nuclear power plants is tritium (^{3}H), a radioactive isotope of hydrogen, composed of one proton and two neutrons. Tritium has a half-life of 12.4 years and as such is radioactive for 248 years. ^{3}H combines readily with oxygen to form tritiated water (T_2O).

Because it is impossible to remove tritium gas or tritiated water via filters, tritium is released continuously from reactors into the air and into lakes, rivers, or seas—depending upon the reactor location. At least 1,360 curies of tritium are released annually from each reactor.[31] (A curie is the amount of radiation equal to the disintegration of 37 billion atoms per second.) Tritium gas is an interesting radioactive material, which is utilized extensively in exit signs, runway signs at airports, and on watch faces. It is very reactive and tends to chemically bind with any material in which it is enclosed.

Tritiated water in particular is scary material. If one is immersed in a cloud of tritiated water on a foggy day near a reactor, it is absorbed straight through the skin. It is also readily absorbed through the lungs and the GI tract. Because tritium is a soft energy beta emitter, meaning that it does not penetrate very far, all the radiation it gives off is readily absorbed by the surrounding cells, hence it is biologically very mutagenic.

There is a vast literature on the biological effects of tritium demonstrating that it causes chromosomal breaks and aberrations. In animal experiments, it has been shown to induce a fivefold increase in ovarian tumors in offspring of exposed parents, while also causing testicular atrophy and shrinkage of the ovaries. It causes decreased brain weight in the exposed offspring and mental retardation with an increased incidence of brain tumors in some animals. Increased perinatal mortality was observed in these experiments as well as a high incidence of stunted and deformed fetuses. (These effects were observed with surprisingly low concentrations of tritium.)[32]

Tritium is also more dangerous when it becomes organically bound in molecules of food.[33] As such it is incorporated into molecules, including DNA within bodily cells. Chronic exposure to contaminated food causes 10% of the tritium to become organically bound within the body where it has a biological half-life of 21 to 550 days—meaning that it can reside in the body from one to twenty-five years.[34]

When tritium is released to the environment, it is taken up by plants and trees, partially incorporating into the ecosystem. Trees constantly transpire water vapor into the air; it has been found that higher concentrations of tritium occur at night at breathing height in a forest that has incorporated tritium from a nearby reactor.[35]

Let's look again at the reactor.

As discussed above, the primary coolant water becomes extremely radioactive over time because the fuel rods leak. The NRC is now allowing nuclear operators to retain uranium fuel in reactors for six years instead of three, lengthening the "burnup" time and substantially increasing the radiation levels in the fuel. The NRC is also allowing a concentration of 4.5% uranium enrichment in the fuel instead of the previously approved maximum of 3.5%. This policy will not only substantially increase the amount of radioactivity produced in the fuel rods, but also subject old reactors to increased power production, which could induce damage in pipes and engineering equipment that have become embrittled and fragile after years of intense radiation exposure. Also, the longer the time that the zirconium fuel cladding is exposed to high levels of radiation, and the higher the radiation levels, the greater the damage to the cladding and subsequent leakage of radioactive materials into the primary coolant.

Radioactive corrosion or activation products that are not the result of uranium fission are also produced, as neutrons bombard the metal piping and the reactor containment. These elements, which are powerfully radioactive, include cobalt 60, iron 55, nickel 63, radioactive manganese, niobium, zinc, and chromium. These materials slough off from the pipes into the primary coolant. Officially called CRUD, it is so intensely radioactive that it poses a severe hazard to maintenance workers and inspectors in certain areas of the reactor.[36]

According to David Lochbaum, a nuclear engineer at the Union of Concerned Scientists, during shutdowns of reactors, the utilities not uncommonly flush out pipes, heat exchangers, etc., to remove highly radioactive CRUD build-up. Some of the CRUD is sent to radioactive waste dumps while some is released to the river, lake, or sea nearest the reactor.[37]

Although the nuclear industry claims it is "emission" free, in fact it is collectively releasing millions of curies annually. Reports documenting gaseous and liquid radioactive releases vary enormously depending upon accidental and larger-than-normal routine releases. The Millstone One reactor in Connecticut alone released a remarkable 2.97 million curies of noble gases in 1975, whereas Nine Mile Point One released 1.3 million curies in 1975. In 1974, the total release from all reactors in the United States was 6.48 million curies, and in 1993 it ranged between 96,600 curies to 214,000 curies.[38] Releases vary according to equipment failure, which is variable and fickle. By contrast, coal plants release some uranium and uranium daughter products in their smoke but very little radiation compared to atomic plants, and certainly no fission products.

The utilities also admit that about 12 gallons of intensely radioactive primary coolant leaks daily into the secondary coolant via the steam generator through breaks in the pipes. Some of these emissions, which occur when the steam is released to the air, are not even monitored.[39] Likewise, about 4,000 gallons of primary coolant water are intentionally released to the environment on a daily basis, while some just leaks out unplanned. Many other emissions are simply not monitored.[40]

Very radioactive primary coolant filters, which often contain intensely carcinogenic plutonium 238, 239, 241, americium, and curium, are shipped to nuclear waste facilities where they will inevitably leak and contaminate water supplies and food chains. But other dangerous elements in the filters are almost certainly present in the primary coolant and escaping in small quantities via the gaseous or liquid effluents into the environment, including: technetium 99 with a 211,100-year half-life, iodine 129 with 15,700,000-year half-life, carbon 14 with a 5,700-year half-life, nickel with a 100.1-year half-life, and plutonium 241 with a

14.29-year half-life. Once in the environment, these carcinogens will bio-concentrate in the food chain, there to enter human bodies![41]

It is important to note that most of the data on radiation releases are not real measurements but are only estimates made by computerized mathematical models based on data generated from operational reactors, field and laboratory tests, and plant-specific design calculations. Hence the nuclear industry is consistently guessing about its radioactive releases and has no real idea what specific isotopes are escaping from its plants. The last document available for public scrutiny that quantified actual releases, not just guesstimates, of radioactive materials from nuclear plants was published by the NRC in 1978. (This was published when reactors were relatively young and plagued with fewer corrosion and maintenance problems.)[42]

RADIOACTIVE WASTE

Quite apart from these routine radioactive releases is the almighty problem of radioactive waste. Each regular 1,000 megawatt nuclear power plant generates 30 tons of extremely potent radioactive waste annually. And even though nuclear power has been operational for nearly fifty years, the nuclear industry has yet to determine how safely to dispose of this deadly material, which remains radioactive for tens of thousands of years. Most nuclear waste is confined in huge cooling pools, euphemistically called "swimming pools" at reactor sites, or in dry storage casks beside the reactor. But there are many other locations in the United States and other countries where huge quantities of reprocessed toxic material and other radioactive waste from nuclear power plants are left unconfined, leaching, leaking, and seeping through soils into aquifers, rivers, lakes, and seas, where it enters and concentrates in the food chains of plants, fish, animals, and humans.[43]

We will now examine several of the precise radioactive materials that the nuclear fission process creates, with their specific health implications for human beings. For simplicity's sake we will consider the properties and medical dangers of only four of these 200 isotopes, giving examples of contamination that has already occurred as they have leaked from their respective reactors.

Plutonium

A typical alpha emitter is plutonium, named after Pluto, the Greek god of hell. Said by its discoverer, Glen Seaborg, to be the most dangerous substance on earth, it is so toxic and carcinogenic that less than one-millionth of a gram if inhaled will cause lung cancer. It is translocated from the lung by white blood cells and deposited in the lymph glands in the middle of the chest where it can mutate a regulatory gene in a white blood cell or lymphocyte causing lymphoma or leukemia. From there it can be solubilized, and, because plutonium resembles iron, it is combined with the iron transporting protein, transferrin, and taken to the bone marrow to be incorporated into the hemoglobin molecule in the red blood cells. Here the alpha particle irradiates bone cells to cause bone cancer and white blood cells made in the bone marrow to cause leukemia. It is stored in the liver where it causes liver cancer, and it is teratogenic, crossing the placenta into the developing embryo.

Plutonium is also stored in the testicle adjacent to the precursor cells, spermatocytes, that form the sperm. Here it will cause mutations in the reproductive genes and increase the incidence of genetic disease in future generations. It also causes testicular cancer. Every male in the Northern Hemisphere has a tiny amount of plutonium in his testicles from radioactive fallout that is still falling on the earth from the upper atmosphere, which was polluted by the atmospheric weapons tests conducted by the United States, the Soviet Union, China, France, and Britain in the 1950s and 1960s.

The half-life of plutonium 239 is 24,400 years, so it remains radioactive for half a million years. Therefore, plutonium lives on to enter and damage reproductive organs for the rest of time, and the genetic mutations it causes are passed on successively to future generations for thousands of years. To give an indication of the length of time involved, it takes up to twenty generations for recessive mutations to come together to express themselves as a specific disease entity, such as cystic fibrosis.

Plutonium is so carcinogenic that the half ton of plutonium released from the Chernobyl meltdown is theoretically enough to kill everyone on earth with lung cancer 1,100 times if it were to be uniformly distributed into the lung of every human being.[44]

Though only 10 pounds of plutonium—a lump the size of a grapefruit—will make an effective atomic bomb, literally hundreds of tons of plutonium are lying around the world, some of it relatively unguarded. The design for an atomic bomb can easily be found on the Internet; some basic materials purchased at the local hardware shop will complete production. The fact that plutonium is a by-product of nuclear power explains why any country that owns a nuclear power plant has access to atomic bomb fuel. Therefore, nuclear power is integral to the ever-growing problem of nuclear proliferation.

However, the United States has historically maintained a strict separation between civilian nuclear power plants and military reactors that produce plutonium for bombs, although they are similar machines. Recently, however, that clear line of separation has been breached, because tritium, which is an integral part of a nuclear weapon, is now being manufactured at the Watts Barr nuclear power plant in Tennessee.

Iodine 131

Radioactive iodine 131, with a half-life of eight days, is a very volatile isotope, meaning that it is usually released from nuclear reactors as a gas, either from routine or accidental emissions. It is both a

beta and a high-energy gamma emitter, and as such it is very carcinogenic. When humans and animals are exposed to this pollutant
in the air, they inhale it into their lungs, where it is absorbed through
the lining of the alveoli or air sacs and enters the blood stream. Iodine 131 also deposits onto the soil near nuclear reactors, where it is
taken up by grass and the leaves of plants and concentrated by orders
of magnitude in grass and vegetables.

When cattle eat this radioactive grass, iodine 131 is concentrated again in their milk. Radioactive iodine enters the human
body in one of two ways—either via the gut when dairy products
from cows eating this grass are consumed or via the lung when radioactive gases are released routinely or accidently into the air from
the reactor. Iodine 131 circulates in the human blood stream and is
avidly absorbed by the thyroid gland at the base of the neck. Children are at special risk from this isotope because their tiny thyroids
avidly absorb iodine from the blood like a sponge.

Strontium 90

Strontium 90 is an isotope released from reactors in small amounts
on a daily basis, mostly in the waste water but sometimes in air. It
is often released in larger quantities when accidents occur at nuclear power plants. It is a beta and gamma emitter with a half-life
of twenty-eight years—radioactively dangerous for 600 years. As a
calcium analogue, strontium 90 mimics calcium in the body. After
release from a nuclear power plant, it lands on the soil, where it is
taken up and concentrated by orders of magnitude in grass, concentrated further in cow and goat milk and in the breasts of lactat-
ing women, where it can induce breast cancer many years later.
Babies who drink this contaminated human breast milk or cows'
milk will be exposed to strontium 90, which enters the gut, is absorbed and carried in the blood stream, and laid down in teeth and
bones, there to induce bone cancer or leukemia years later.

Cesium 137

Cesium 137 is an isotope with a half-life of thirty years, radioactive for 600 years. As a potassium analogue, it is present in every cell of the body. Cesium 137 tends to concentrate in animal muscle and fish, and it deposits in human muscles where it irradiates muscle cells and other nearby organs. It is a dangerous beta and high-energy gamma emitter and is very carcinogenic. An old, dirty reactor at Brookhaven National Labs in the middle of Long Island in the 1970s and 1980s released large amounts of radiation for many years,[45] and an epidemic of a very rare form of cancer called rhabdomyosarcoma appeared in children living near that reactor in the 1980s. This very malignant muscle cancer could be caused by exposure to cesium 137.[45a]

Dr. John Gofman, the discoverer of uranium 233, estimates that if 400 reactors operated for twenty-five years at 99.9% perfect cesium containment, cesium loss over this period would be equivalent to sixteen Chernobyl accidents.[46]

NUCLEAR ACCIDENTS

Abnormal releases of small or large quantities of radiation at nuclear power plants occur not infrequently and are referred to by the nuclear industry as "incidents." These "incidents" occur because of human or mechanical error or because the operator at the reactor has purposefully decided to vent radioactive gases to get rid of them.

Several incidents have had catastrophic ramifications. A meltdown occurred at the Three Mile Island reactor in the United States and a massive power excursion erupted at the Chernobyl nuclear power plant in Russia. These were both induced by human error and fallibility. Because the reactors around the world are aging and suffering from cumulative metal fatigue induced by high radiation exposure, there is a high probability that another meltdown will occur in the near future.

There have been several near-misses at large nuclear power plants in the United States over the years. One took place at the Browns Ferry reactor in Alabama in 1975, when two electricians using candles to check for air leaks accidentally ignited highly combustible polyurethane foam that was used as sealant. The fire rapidly spread to plastic cables that surrounded other cables that controlled the operation of the reactor and emergency core cooling system. The fire raged in the bowels of the plant for seven-and-a-half hours, severing thousands of cables and debilitating most of the control systems and the emergency core cooling system. The water level in the reactor core dropped sharply and was restored only when workers resorted to equipment that was not designed for emergency cooling systems.[47] A more recent "near-miss" occurred at the Davis-Besse plant in December 2001; this accident is described in the next chapter.

Three Mile Island

Before Three Mile Island melted down, the nuclear industry used to say that the chance of a meltdown occurring was the same as that of a person being hit by a bolt of lightning in a parking lot.

Beginning at 4 A.M. on March 28, 1979, lightning struck. A meltdown at the Three Mile Island nuclear power plant in Pennsylvania was triggered when a mechanical failure and an automatic shutdown of the main feedwater pumps in the secondary coolant system closed some valves, causing water in the primary coolant system covering the radioactive core to overheat. This quickly cascaded into a series of automated events and human misinterpretations, which caused the reactor core of 100 tons of uranium to overheat and to melt. Throughout the accident, highly radioactive cooling water was being pumped through a valve onto

the floor of the reactor and thence into a tank in an adjacent auxiliary building where large quantities of radioactive gases were
vented from a leaking valve into the external atmosphere.

Warm weather at the time of the leak compounded the crisis,
with low winds and a cold upper air mass preventing the warm air
from rising, producing ideal conditions for trapping the radioactive
emissions.[48]

We know for a fact that large amounts of radioactivity escaped from the Three Mile Island accident. But, the nuclear industry and the government did not collect release estimates for
specific isotopes,[49] and to this day, there is no available information
about which specific isotopes escaped nor the actual quantity of
radiation that was released.[50] The gamma radiation monitor on the
auxiliary building where all the radiation was released was not
designed to measure such high concentrations of radiation, so it
went off scale very early in the accident, an emergency which continued over several days. Thus, the only estimates of radiation release
were made by extrapolating data obtained from gamma radiation
monitors—thermoluscent dosimeters (TLDs)—which were located
hundreds of feet from the stack low down on the fence line that surrounded the reactor. Of the twenty TLDs (that only measure gamma
radiation, not beta, which was three to five times the gamma dose),
only two were anywhere near the "hot" passing cloud, hence it is
impossible to judge the dose to thousands of people from only two
readings.[51] Most of the radioactive plumes would have been lofted
into the air well above these monitors, so only small increments of
radiation in the gaseous plumes could possibly have been measured.
Measurements of noble gas were not commenced until April 5, some
eight days after the meltdown first began. No alpha or beta radiation
was ever measured. It is known that radioactive emissions from
Three Mile Island travelled long distances. For instance, xenon 133
was measured in Albany, New York, at the end of March and early
April 1979, 375 kilometers from the reactor.[52]

Radiation releases and dose estimates were therefore determined using extremely inadequate data. The nuclear industry estimated that 13 to 17 curies (1 curie is the amount of radiation equal to the disintegration of 37 billion alpha or beta particles per second) of radioactive iodine escaped, plus 2.4 to 13 million curies of the noble gases krypton, xenon, and argon.[53] But, as the former chairman of the Nuclear Regulatory Commission (NRC), a government-appointed body that oversees the regulation of nuclear power, Joseph Hendrie was quoted at the time as saying, "We are operating almost totally in the blind, [Governor Thornburgh's] information is ambiguous, mine is nonexistent and—I don't know—it's like a couple of blind men staggering around making decisions."[54]

However, based on measurements of radioactive iodine in animals nearby, experts felt the nuclear industry's estimates were grossly understated. Also the March 24, 1982 notes of Dr. Karl Morgan, estimated that 45 million curies of noble gases were released and 64,000 curies of radioactive iodine were released and that the thyroid dose to the population was at least 100 times that of the NRC estimate.[55] Dr. Morgan was a highly respected health physicist known as the "Father of health physics."

Dr. Carl Johnson M.D., M.P.H., an expert in radiation related diseases, estimated that because the fuel melted, many other elements almost certainly escaped from the reactor core, including plutonium, strontium, and americium. When he asked the NRC and the DOE to do a survey to look for these elements in the respirable dust around Three Mile Island after the accident, they refused.[56]

It is known that on day three of the accident, 172,000 cubic feet of high-level radioactive water were released into the Susquehanna River by the utility without NRC permission, an event unheard of in the history of the nuclear industry. The Susquehanna River drains into Chesapeake Bay, a major fishing location.[57] This water contained high concentrations of many dangerous, long-lived isotopes, which would then have been avidly bio-concentrated by fish, lobsters, and crabs over a period of weeks, months, and years. The public, however, was not notified about this danger, as they were not notified about many aspects of this accident.

Large quantities of radioactive krypton 85 were purposefully vented from the damaged reactor in June 1980, exposing even more people to radioactive contamination.[58] And in November 1990, 2.3 million gallons of radioactive water containing tritium was also purposefully evaporated from the damaged reactor building, exposing many people in the vicinity to dangerous radioactive elements.[59]

During the first two days of the accident, pandemonium ensued as 5% to 6% of the people who lived within five miles of the plant fled. On March 30, two days after the accident, Governor Thornburgh ordered the evacuation of pregnant women and children from the five-mile zone.[60] One hundred and forty-four thousand people packed up and fled, jamming the highways, with babies bundled in blankets, children with scarves wrapped across their faces to limit their exposure to radiation, and pregnant women in sheer panic.[61] I was in Harrisburg, Pennsylvania, a week after the accident to explain the effects of radiation to thousands of frightened residents in the gymnasium of the high school when it was reported to me that local physicians fled with their families, leaving their patients in the hospitals to fend for themselves.

Hundreds of local people reported a variety of symptoms and signs that were similar to the symptoms reported almost a decade later by residents of Pripet, the town adjacent to Chernobyl where another nuclear meltdown occurred and the release of radiation was much greater than that at Three Mile Island.[62] These symptoms included nausea, vomiting, diarrhea, bleeding from the nose, a metallic taste in the mouth, hair loss, and a red skin rash. These are the typical signs and symptoms of acute radiation sickness, which manifest when people are exposed to whole-body doses of radiation around 100 rads—a high level of exposure. This dose kills the actively dividing cells of the body—hair, gut, and blood—a situation that induces these symptoms. The people near Three Mile Island also reported deaths of farm animals and pets.[63]

Dr. Gordon McLeod, Pennsylvania health commissioner at the time of the accident, noted that the number of babies born with hypothyroidism increased from nine in the nine months prior to the accident to twenty in the nine months following the accident, and he postulated that this was because the thyroid gland was affected by the large quantities of iodine 131 that escaped during the accident. Dr. McLeod's finding indicates that some people who were exposed to iodine 131 as babies and children may well be developing thyroid cancer as they are in the exposed population from Chernobyl, but if nobody investigates the situation epidemiologically these patients will not be identified. McLeod was fired by Governor Richard Thornburgh just six months after he took office.[64]

A Food and Drug Administration document dated April 6, 1979, analyzed milk collected on April 4, 1979 from many farms in the area surrounding the Three Mile Island reactor. Fifteen of the samples showed elevated levels of iodine 131 and twelve showed elevated levels of cesium 137.[65] The farms were located at all four sectors of the compass, meaning the radioactive plume moved 360 degrees from the day of the accident to seven days thereafter. It's also interesting that the farms whose milk tested positive were varying distances from the reactor—from 150 miles north to 9 miles south, 15 miles west to 13 miles east. Hershey Chocolate factory is located 13 miles from Three Mile Island in the richest dairying area of the United States. At the time of the accident most of Hershey's milk supply came from Pennsylvania.

A memo written on April 11 to W.J. Crook, of Hershey's Science and Technology Department by C.J. Crowell, the Quality Assurance Manager of Hershey's, in response to a discussion between the two men the previous day is significant. One statement in this document said that they continued to check the liquid milk inside the five mile zone adjacent to the plant and that "no detectable radiation has been found since a few days after the accident," which appears inconsistent with the information in the previously mentioned FDA report where radiation was found in the milk up to one week post accident many miles outside the five mile zone.[66]

However a confidential memo from Hershey states that between the dates of April 2 to April 20, 12,270,000 pounds of milk were powdered, instead of 6,095,000 pounds that would normally have been converted to powder during that time. When milk is powdered it is preserved in a usable form until the radioactive iodine decays. However this technique does not obviate cesium 137 in the milk, which lasts 600 years, nor other long-lived radio-isotopes should they also be present in the contaminated milk. Hershey was obviously concerned about their milk supply; however, they repeated in this document that no detectable radiation was found in this milk "since a few days after the accident."

But another study also performed on milk on March 30, 1979 by the Pennsylvania State University, College of Engineering, which was sent by K.K.S. Pillay to Dr. Carl. Y. Wong, Group Leader of Product Research at Hershey Foods Corporation, found 3,000 picocuries/liter in milk from farms located 12 and 15 miles from the reactor, 3,500 picocuries/liter from a farm 7 miles away, 4,000 picocuries/liter from another farm 16 miles away, and three calculations from unidentified farms measuring 6,000 picocuries/liter, 8,500 picocuries/liter to 21,500 picocuries/liter.[67] Some of these levels are very high. If a one-year-old child drank a liter of milk containing 21,300 picocuries/liter, she would receive a dose of about 0.3 rems to her thyroid, which could result in thyroid cancer years later. If she consumed more than one liter of contaminated milk the dose would increase accordingly.

However a quote from Thomas Gerusky, the director of Radiological Health from the Pennsylvania Department of Environmental

Resources (DER) in the April 8 edition of the *Harrisburg Sunday Patriot News* said that, "If we ever found a thousand picocuries we would take action." These measurements from Pennsylvania State University indicate that action should have been taken.[68]

The question is, was the DER informed about the high levels of radiation in the milk, or were they hidden? If on the other hand the DER was informed, why did they do nothing?

The cows were out at pasture on the day of the accident because it was early spring, but on March 29 they were removed from the fields and fed on silage, which would not have been contaminated. The cows were therefore not subject to excessive radiation absorption through their GI tracts. Therefore radiation must have entered through their lungs, absorbed from highly contaminated air.

If the cows were contaminated, so too were humans. Why then were the cows tested for contamination but not people?

This question remains pertinent twenty-six years post meltdown. If radioiodine and cesium 137 escaped, so too did various quantities of strontium 90, plutonium, americium, and other extremely dangerous and long-lived materials. What are the ground measurements of these elements on the land where the cows graze that produce milk for Hershey's chocolates, if indeed they were ever taken or analyzed? Why has this data never been released?

Subsequently and strangely, there has been a deficit of studies performed on the medical outcomes of this accident and a plethora of studies relating people's symptoms to stress. Two medical papers emerged from Columbia University investigators, which reported a positive association between radiation exposures and increased incidences of non-Hodgkin's lymphoma, lung cancer, all other cancers combined, plus childhood leukemia. However, these findings were statistically insignificant because of the small overall numbers of cases (fifty-four), and the team used estimated radiation exposures performed by the government and nuclear industry, which were artificially small. The Columbia team decided that their find-ings were unrelated to radiation exposure because the doses had been too low to initiate the increase in the malignant diseases that they found. They also postulated that increased cancer rates were caused by stress![69]

However, another study later performed by Steve Wing and others found positive relationships between accident dose estimates and the increased incidence of cancer that was reported by the Columbia group.[70] They used the same dose estimates as did the Columbia team, but they did not make the same assumption that the absolute radiation level to which the public was exposed was below the background levels of radiation.

The official health studies were paid for by the TMI Public Health Fund, which was set up by the nuclear industry and funded by industry payments, which also settled property damage suits. At no stage did the nuclear industry confer with or obtain evidence from citizens who believed that they had been impacted by the ac-cident.[71] These people were excluded when the questions were formulated; they were excluded from participation in the study de-sign, interpretation, and analyses; and they were not told of the re-sults. To add insult to injury, those people who dared to testify about their experiences and physical symptoms were often sub-jected to ridicule at hearings.[72]

The Three Mile Island case eventually ended up in court, when approximately 2,000 residents claimed that the radioactive releases from the meltdown had been much larger than those offi-cially proclaimed by the nuclear industry and government officials. After several dismissals and appeals, however, the plaintiffs decided that they could no longer afford to continue and had to settle.[73]

Dr. Wing, a famed epidemiologist who represented the plaintiffs, noted for the record his impression that the industry's image and lia-bility were more important than the accuracy of data and full disclo-sure. Wing pointed out that, historically, disputes between industry, governments, and community are always unequal and unfair because people who have been damaged by irresponsible industries almost never have the expertise or funding to conduct their own studies.[74]

Surprisingly, the cancer incidence in the exposed population was studied only through the years 1981–1985, several short years post-accident.[75] There have been virtually no further epidemiological studies performed since that time, even though the latent period of carcinogenesis is two to sixty years and even though the long-lived isotopes strontium 90, cesium 137, and others almost certainly escaped during the accident.

It is imperative that further studies be implemented on the exposed Three Mile Island population. In a relevant precedent, for many years scientists insisted that the fallout from the U.S. aboveground nuclear weapons testing during the 1950s and 1960s had produced no health impacts. In 1997, when studies were finally performed, the National Cancer Institute (NCI) estimated that as many as 212,000 Americans had developed or would develop thyroid cancer from the radioactive iodine released from the tests.[76] And even then this study was inadequate because the NCI did not estimate different types of cancer that would have been induced by the many other radioactive elements such as strontium 90, cesium 137, and plutonium that were released from these aboveground nuclear tests.

In 1991, Dr. Karl Morgan, the founder of the discipline of health physics, looking retrospectively at the track record of that field, wrote that health physics was intended "to be a science and profession to protect radiation workers and members of the public from exposure to ionizing radiation. It succeeded to some extent in this objective but during the past decade in the United States, it has reverted to an organization primarily to protect the nuclear industry from liability resulting from radiation exposure."[77]

There were two reactors at Three Mile Island. One is still in operation generating electricity. The melted core of the damaged reactor has been dismantled. It took some eleven years to clean up the melted and fragmented fuel rods, which were intensely thermally and radioactively hot. Ninety-nine percent of this material was sent by truck to the Hanford reservation in Washington State and to the Idaho National Engineering Labs in Idaho Falls. The reactor building itself remains intensely radioactive, significantly more so than a reactor building at the end of its forty years of operation.[78]

A lawsuit initiated in 1992 by Eric Epstein against General Public Utilities, then owner of Three Mile Island, resulted in the establishment of a state-of-the-art monitoring system around Three Mile Island. Gamma monitoring equipment, which itself is continuously monitored, has been deployed in sixteen locations within three miles of Three Mile Island. During the years 2003–2005, this monitoring system was enhanced by the addition of five state-of-the-art radiation monitors, which feed information into the central control system at Penn State University.[79] No further cancer studies have been implemented in the exposed population, although a 2002 report issued by the National Cancer Institute and the Centers for Disease Control and Prevention found that Pennsylvania had the seventh-highest cancer incidence in the nation.[80]

In February 1985, however, $3.9 million dollars were paid out in settlements to people who had developed diseases related or unrelated to radiation by the insurance company representing General Public Utilities Corp. and Metropolitan Edison Co., the owners of Three Mile Island. The claimants were told there could be no further claims by them for liability and in exchange for the compensations paid they agreed not to discuss the settlements.[81]

Chernobyl

Before the Chernobyl meltdown, the nuclear industry assumed that, in the event of an accident at a nuclear power plant, only a tiny percentage of the radioactive inventory of the reactor core would escape from the containment into the environment. On April 26, 1986, when Unit Four of the Chernobyl nuclear power plant exploded, however, almost all the contents of the deadly radioactive fission products were spewed into the environment.[82] This medical catastrophe will continue to plague much of Russia, Belarus, the Ukraine, and Europe for the rest of time.

However, in 2005, the International Atomic Energy Agency (IAEA) produced a United Nations report on Chernobyl, claiming that only fifty-six people had died as a result of the accident. The IAEA has a conflict of interest when it comes to monitoring the health consequences of radiation because, in 1959, the IAEA signed a somewhat diabolical agreement with the World Health Organization (WHO), preventing WHO from researching health consequences emanating from atomic, military, and civilian use of the atom, even preventing them from issuing warnings to exposed populations. Dr. Michael Fernex, formerly on the faculty of the University of Basel, who worked with the WHO, said in 2004, "Six years ago we tried to have a conference. The proceedings were never published. This is because in this matter the organizations at the UN are subordinate to the IAEA. . . . Since 1986, the WHO did nothing about studying Chernobyl. It is a pity. The interdiction to publish which fell upon the WHO conference came from the IAEA. The IAEA blocked the proceedings; the truth would have been a disaster for the nuclear industry."[83]

So in order to prevent a disaster befalling the nuclear industry, the magnitude of the *true* disaster is deliberately being obfuscated.

These are some of the medical and ecological consequences of Chernobyl that we know today:

- Of the 650,000 people called "liquidators" involved in the immediate cleanup, 5,000 to 10,000 of them are known to have died prematurely.[84]

- Large areas of the breadbaskets of the Ukraine and Byelo-Russia became heavily contaminated and will remain so for thousands of years. In all, 20% of the land area of Belarus, 8% of the Ukraine, and 0.5% to 1% of Russia—100,000 square miles—were contaminated. In total, this area is equivalent to the state of Kentucky or of Scotland and Ireland combined. Five million people live in these areas, over 1 million of whom are children, who are inordinately sensitive to radiation. The incidence of cancer among this population has increased. Many of the genetic abnormalities and diseases caused by this accident are generations away and will not be seen by anyone alive today.
- Heavy radioactive fallout occurred over Austria, Bulgaria, Czechoslovakia, Finland, France, East and West Germany, Hungary, Italy, Norway, Poland, Romania, Sweden, Switzerland, Turkey, Britain, the Baltic States, and Yugoslavia. Small amounts also landed on Canada, the United States, and all other countries in the Northern Hemisphere.[85] Because cesium 137 and other isotopes such as strontium 90 and plutonium 239 have such long half-lives, some of the food in Europe will be radioactive for hundreds of years, depending upon the hot spots that were contaminated when the radiation fell to the earth as rain.
- In Britain, twenty-eight years post-accident and 1,500 miles from the crippled reactor, 382 farms containing 226,500 sheep are severely restricted because the levels of cesium 137 in the meat are too high. Before the sheep are sold for meat, they must be transferred to other less radioactive grazing sites so that their levels of cesium decrease before sale.[86] Meanwhile, people in Britain are still eating low levels of cesium in their meat.
- In the south of Germany, very high levels of cesium in the soil persist; hunters are compensated for catching contaminated animals, and many mushrooms and wild berries are still too radioactive to eat.[87]

- The French government initially insisted that the radioactive fallout stopped exactly at the French border. Recent documents reveal, however, that the government knew that radioactivity in France surpassed all safety levels at the time of the accident.[88] Other European countries ruled that fresh vegetables and dairy products could not be sold for several months and that children were not to play outside for a similar short time span, but the French government denied that France was affected. Only now do they admit that cesium 137 in some parts of France is as high as some extremely contaminated areas in Belarus, the Ukraine, and Russia. A country that loves its food, mushrooms, and wild boar shows very high levels of contamination, mainly in the form of cesium 137.[89] Perhaps the fact that France has fifty-eight nuclear reactors and derives 80% of its electricity from nuclear power is related to the government's cover-up.[90]

The reindeer as far away as Scandinavia were contaminated with cesium after the Chernobyl meltdown, because the lichen in the Arctic Circle avidly concentrated the cesium as it landed on them from the fallout. (I visited Sweden just after the accident and stupidly ate some reindeer meat!) Signs in Bavarian forests warn people not to eat the mushrooms—this is because they are very efficient concentrators of radiation, particularly cesium 137.

In all my years of pediatric practice, I have never seen a child with thyroid cancer, because childhood incidence is so extremely rare. Yet in Belarus near Chernobyl from 1986 to 2001, 8,358 cases of thyroid cancer occurred, 716 in children, 342 in adolescents, and 7,300 in adults.[91] The situation post-Chernobyl is a medical emergency, unique in the history of pediatrics. Most of those affected have had their thyroids surgically removed, but a person cannot survive without the hormones produced by the thyroid gland, so these children and adults are dependent upon receiving thyroid replacement tablets every day for the rest of their lives. Should some catastrophic situation such as a war impede their drug supply, they will die.

Chernobyl also impacted the daily lives of 400,000 people who resided in the most contaminated areas of the Ukraine, Belarus, and Russia. Because of the accident, they were forced to leave their homes, their past, their friends and their communities forever. Many were relocated to other areas (often to find that the land was just as radioactive as the homes and gardens that they vacated). They now live with the fact that they and their children are forever contaminated and could develop cancer or produce a new generation of children with severe birth defects.[92]

Although the food in many parts of Europe is still relatively radioactive as evidenced in the data presented in this chapter, this terrible problem is rarely mentioned in the media or in daily conversation. In a form of psychic numbing, people continue to live their lives as if all were well, and the nuclear power industry continues to broadcast the myth that its product is clean and green.

In 1994, the United Nations Office for the Coordination of Human Affairs made a tragic statement of remembrance, almost like statements made to memorialize wars:

> Eighteen years ago today, nearly 8.4 million people in Belarus, Ukraine and Russia were exposed to radiation. Some 150,000 square kilometres, an area half the size of Italy, were contaminated. Agricultural areas covering nearly 52,000 sq km, which is more than the size of Denmark, were ruined. Nearly 400,000 people were resettled but millions continue to live in an environment where continued residual exposure created a range of adverse effects.
>
> Now, roughly 6 million people live in affected areas. Economies in the region have stagnated, with the three countries directly affected spending billions of dollars to cope with the lingering effects of the Chernobyl disaster.

Chronic health problems, especially among children, are rampant.[93]

Eighteen years after the accident, 70% to 90% of the cesium 137, 40% to 60% of the strontium, and up to 95% of the plutonium and its alpha-emitting relatives remain in the upper root–inhabiting layers of the soil in Belarus, Ukraine, Russia, and parts of Europe.

In 2001, the United Nations Development Program-United Nations Children's Fund (UNDP-UNICEF) mission summarized:

The health and wellbeing of populations in the affected regions is generally very depressed. . . . Life expectancy for men in Belarus, Russia and Ukraine, for example, is some ten years less than in Sri Lanka, which is one of the twenty poorest countries in the world and is in the middle of a long drawn out war. . . . Cardiovascular disease and trauma (accidents and poisonings) are the two most common causes of death followed by cancer (this situation is not confined to the Chernobyl affected regions). . . . The health situation encountered in the populations living in the affected territories is thus a complex product of inputs ranging from radiation induced disease, through endemic disease, poverty, poor living conditions, primitive medical services, poor diet, and the psychological consequences of living with a situation that was frightening, poorly understood, and over which there seemed little control.[94]

Thus, the extreme degree of dislocation, fear, and anxiety precipitated by the nuclear disaster also aggravates and potentiates other diseases.

A recent study from Sweden showed an increase of 849 cancers up to the year 1996 as a result of Chernobyl.[95] This is the first study outside Russia, Ukraine, and Belarus to show an effect. It will be the first of many, because the time frame of ten years is relatively short for the incubation of cancer and because other countries have yet to study their affected populations. There are now claims surfacing in France that people are suffering from thyroid cancer that may be related to the Chernobyl fallout.[96]

To put a final terrifying coda to this story, the Chernobyl reactor's radioactive sarcophagus, which was hastily constructed to cover some 20 tons of melted fuel and radioactive dust at the site of the damaged reactor, is disintegrating and cracking and is not expected to remain intact for many more years.[97] If it collapses, it will release huge quantities of radiation that will again be swept across the Ukraine, Belarus, Russia, and parts of Europe, depending upon the wind direction.

The accident is not over.

4

Accidental and Terrorist-Induced Nuclear Meltdowns

Nuclear power plants are vulnerable to many events that could lead to meltdowns, including human and mechanical errors; impacts from climate change, global warming, and earthquakes; and, we now know, terrorist attacks.

MECHANICAL AND HUMAN ERROR

A recent near miss at the Davis-Besse reactor twenty-one miles southeast of Toledo, Ohio, caused the reactor to come within days or weeks of a major catastrophe. To save money, the owner, First Energy, had persuaded the NRC to delay inspections of a vital safety component beyond the due date of December 31, 2001.[1]

When the reactor was eventually shut down in February 2002 for an early refuelling outage, to their horror the inspectors came upon a cavity, where corrosion had eaten its way through six inches of carbon steel on top of the six-and-a-half-inch reactor pressure vessel. Less than half-an-inch of stainless steel liner of the reactor remained to separate the pressurized internal radioactive environment from the reactor containment building. This stainless steel liner was bulging outward but luckily had not ruptured.

How did this hole occur? Reactor coolant water, which contains boron, had escaped through cracks and flanges around the control rod drive mechanisms, to drip and crystallize on the exterior surface of the carbon steel reactor head. Because boric acid is very corrosive, over time the boron acid ate its way through the carbon steel, creating a hole four-by-five-by-six inches. As a result, a wide surface area of thin stainless steel was exposed to extremely high pressures beneath it.

Had this steel ruptured, reactor coolant water would almost certainly have been released as a high-pressure jet stream. This energetic jet stream would almost certainly have damaged reactor safety equipment located immediately above it. It also could potentially have created a shock wave powerful enough to break control rods that were already cracked. These would then have been ejected as missiles, which would have created chaos among safety equipment, including specific control rods that close down the reactor. This cascade of events could have initiated a core meltdown.[2]

But that is not the end of the story. Later in 2002, the owner of the Davis-Besse plant informed the Nuclear Regulatory Commission that when the hole was discovered, a large amount of debris was simultaneously discovered in the containment building, which could potentially have blocked the emergency sump intake screen, rendering the sump inoperable in the event of a loss-of-coolant accident. This would have caused both the Emergency Core Cooling Systems and the Containment Spray systems to be rendered inoperable, because both require emergency sump suction during the recirculation phase. So exactly when the Emergency Core Cooling system might have been needed, it was inoperable, as was the filter to remove escaping radioactive iodine in the event of a meltdown.[3]

A senior NRC manager who had been involved in the decision to allow Davis-Besse to continue operation said he felt the agency's hands were tied: "We can argue this, but this agency does not take precipitous action to shut down a nuclear power plant because we have a suspicion of something without enough evidence to warrant it. . . . If we were in the same situation again, we'd probably make the same decision" to allow them to operate until February 16.[4]

The Davis-Besse emergency is one of many, if not all so serious, that confront the nuclear industry every year. Statistically speaking, an accidental meltdown *is almost a certainty* sooner or later in one of the 438[5] nuclear power plants located in thirty-three countries around the world. Human error, compromise, laziness, and greed are implicit in the affairs of men; when these attributes are applied to the generation of atomic energy, the results can be catastrophic.

AGING REACTORS

Even though today's reactors were designed for a forty-year life span, the NRC, acceding to industry pressure, is currently approving twenty-year extensions to the original forty-year licenses for nuclear power plants.[6] But as David Lochbaum, a nuclear engineer from the Union of Concerned Scientists, points out, nuclear power plants are like people: they have numerous problems in their infancy and youth, they operate relatively smoothly in early-to-middle life, and they start to show signs of stress and manifest pathology as they age.[7]

Every U.S. nuclear power plant is moving into the old-age cycle, and the number of near-misses is increasing. In a thirteen-month period from March 7, 2000, to April 2, 2001, eight nuclear power plants were forced to shut down because of potentially serious equipment failures associated with aging of their mechanical parts—one shut down on average every sixty days. The NRC aging-management programs are thus failing to head off the equipment failures these programs are designed to prevent.[8]

Specific examples include the Oconee Unit 3 in South Carolina where, on February 19, 2001, boric acid was found on the exterior surface of the reactor vessel head around two control rod drive mechanism (CRDM) nozzles. Further investigation found circumferential cracks that went right through the reactor vessel head wall above the weld areas where the nozzles were attached to the reactor vessel head. On January 9, 2002, at Quad Cities Unit 1, in Illinois, operators shut down the reactor when one of the jet pumps inside the reactor vessel had failed. Further investigation found that the hold-down beam for jet pump 20 had cracked apart, and pieces had damaged the impeller of the recirculation pump, causing it to shut off. On October 7, 2000, workers found boric acid on the containment floor at the Summer nuclear plant in South Carolina, and this finding led to the discovery of a through-wall crack where a major pipe was welded to the reactor vessel nozzle. This area had been previously inspected in 1993, but the crack was missed because an air gap between the pipe weld area and the inspection detector had created a "noisy" output, which masked the indications of the crack. And on February 15, 2000, a steam generator at Indian Point Unit 2 plant in New York, thirty-five miles from the center of Manhattan, released 19,197 gallons of intensely radioactive water from the primary coolant into the atmosphere.[9] The owner of the plant had detected indications of degradation during steam generator inspections in 1997 but had failed to do anything about the problem.[10]

As the Union of Concerned Scientists notes, "These examples illustrate two fundamental flaws in current aging management programs:

1. looking in the wrong spots with the right inspection techniques (as happened at the Oconee and Quad Cities plants)
2. looking in the right spots with the wrong inspection techniques (as happened with the Summer and Indian Point plants)."[11]

Although aging management programs should find these problems before they become self revealing, they have not.

Serious problems associated with metal fatigue and aging in reactors occur within the steam generators, with many experiencing cracked and broken steam generator tubes. Steam generator tubes are very important because they constitute more than 50% of the barrier between the primary coolant and secondary coolant in a pressurized water reactor. And there is no containment vessel over the steam generator or any other mechanism to prevent fission products from entering the environment. Most importantly, leakage of primary coolant from the steam generator in the event of a rupture could seriously deplete the primary coolant, leading to a meltdown.[12] Yet the NRC for the last ten years, although they have been aware of this problem, have allowed many nuclear power plants to operate with thousands of cracked steam generator tubes.[13]

The NRC is also allowing plant owners to test their emergency equipment less frequently than required by law and to operate degraded equipment. Furthermore, the NRC does not make public its risk assessment studies on nuclear power plants, even though by law it is obliged to do so. David Lochbaum says this "agency continues to make regulatory decisions affecting the lives of millions of Americans in a vacuum."[14]

It is impossible to know whether an aging plant can operate safely for another twenty years. As Lochbaum points out, "There needs to be strong aging management programs at all reactors to ensure failures are found before it is too late." The best way to prevent recurrent problems at aging reactors, Lochbaum argues, "would be for the NRC to suspend the issuance of license renewals until the nuclear industry has demonstrated that it takes plant safety seriously. Plant owners will continue to follow lax aging management programs and allow failures to reveal themselves unless the NRC imposes stronger standards." Construction of new reactors is enormously expensive, whereas reactors that have been operating for many years are relatively inexpensive to maintain, and the profits to their owners are large.[15] Right now, as Lochbaum points out, the owners have few economic incentives to retire their aging plants voluntarily.[16]

GLOBAL WARMING

In August 2003, France experienced such a severe heat wave that 14,000 people died. Many of these were old people, who lived in non-air-conditioned apartments. The hot weather and lack of rainfall severely reduced supplies of cold river water, and when the river levels fell, the French power company, Electricite de France, resorted to cooling its nuclear power plants by hosing down their outsides with garden sprinklers supplied by reservoirs.[17] Eventually, because the reactors were operating at higher-than-normal temperatures, the company was permitted to discharge secondary cooling water into the rivers where the plants are located at temperatures that destroy aquatic life.[18]

The impact of global warming presents a very serious situation for nuclear energy for several reasons:

- Global warming can induce unpredicted and extreme weather events that could heat up the rivers and lakes from which nuclear power plants extract their cooling water. An adequate supply of water itself may also cease to exist as drought conditions take over.
- Nuclear power plants were designed thirty to forty years ago, before the advent of global warming was considered, and nuclear regulations are not cognizant of global warming. Current regulators tend to minimize the risk. For instance,

Scott Burnell, a spokesman at NRC, said new power plant guidelines were unnecessary, because "global warming occurs on such a slow scale that we would be able to deal with any changes at an operational level as opposed to a policy level."[19]

- The extrusion of very hot water from reactors poses an enormous risk to aquatic life already feeling the stresses of global warming. Five or six years of water temperatures lethal for salmon could induce extinction in certain areas.

- As temperatures rise, so the human need for more air-conditioning and more electricity to drive air conditioners increases. This then becomes a vicious cycle: more heat, more air conditioners, more electricity, more CO_2 production, and more stress upon nuclear power plants and their associated environment.

TSUNAMIS AND EARTHQUAKES

Many nuclear power plants around the world are susceptible to the effects of tsunamis because they are located by the sea, from whence they extract their cooling water. In 2004, a tsunami struck a reactor in India; although it did not induce a major accident, it did cause a degree of damage. The height of the tsunami that originated off the coast of Thailand in December 2004 was a massive ninety-eight feet. In California, the reactor at Humbolt Bay is only twelve feet above sea level. Although it is now closed down, it still contains much of its highly radioactive fuel.

Diablo Canyon and San Onofre, both also in California, are similarly vulnerable to tsunamis and are also located adjacent to earthquake faults. Although nuclear reactors are designed to withstand serious earthquakes, a quake large enough could trigger a very severe accident. This is true as well for many of the reactors in Japan, a country riven with earthquake faults, and for many other earthquake-prone countries. Pacific Gas and Electric, the company that owns the Humboldt Bay reactor, decided to close it down because of the risk that an earthquake could trigger a severe accident.[20]

TERRORISTS

The report of the 9/11 Commission revealed that al Qaeda had considered plans to attack nuclear power plants. The commission thought that scenario was unlikely, however, because of their mistaken belief that the airspace around nuclear power plants was "restricted" and that planes violating that air space would be shot down before impact.[21]

They were wrong. No-fly zones around reactors do not exist on a standard basis, even today; they are imposed only at times of heightened threat. No surface-to-air missiles are deployed around nuclear power plants. (Many such missiles are deployed around Washington, D.C., however, since 9/11.[22] Evidently, politicians have decided that it is more important to protect their own lives than those of the millions of people who could die lingering, painful deaths after a terrorist-induced nuclear meltdown.)

According to John Large, a UK consulting engineer, in an article in *Global Health Watch*, "Nuclear power plants are almost totally ill-prepared for a terrorist attack from the air" because nuclear reactors were designed and constructed more than fifty years ago, well before the large airplanes in common use today were ever conceived.[23] According to Large, a full-sized passenger plane travelling at great speed with a full load of fuel could significantly damage a nuclear reactor, while injecting large quantities of burning fuel into vulnerable areas of the building. This, in turn, could induce enough damage that a meltdown would occur, leading to the release of large quantities of radiation.[24]

Most nuclear reactors are not required by the NRC to be able to withstand attacks from planes or boats.[25] Large points out that designs

of relevant nuclear power plants are easy to obtain in the open literature,[26] and he says that there are no practical measures to take to
ensure reactors will not be severely damaged.[27] Others, however, have
recommended that the vulnerable aspects of a nuclear power plant
could be protected by a series of steel beams set vertically in deep
concrete foundations connected with bracing beams, a web of high-
strength cables, wires, and netting linking the vertical beams to form a
protective screen. This so-called Beamhenge would act to slow down
an attacking aircraft, fragmenting it into smaller pieces and dispersing
the mass of jet fuel, thereby protecting the vulnerable containment
vessel, the spent fuel pool, and other vital pieces of equipment. The
NRC has yet to implement any such protective measure.[28]

The external electricity supply to reactors and the emergency
diesel generators upon which the safe operation of a nuclear reactor depends are also susceptible to terrorist attack, as is the intake of
cooling water from the nearby sea, river, or lake.[29]

SECURITY

Time magazine recently examined the degree of security available
at nuclear power plants post 9/11. Although security at civilian
airports has been enormously improved, security at nuclear power
plants is virtually unchanged, even though these facilities constitute
potential weapons of mass destruction and, as such, are inviting
targets for terrorists. In truth, terrorists do not need their own
weapons of mass destruction, as such weapons are conveniently deployed all over the world next to large and strategically important
populations.

Time magazine opens its article with an attack scenario that
goes like this:

> The first hint of trouble probably would be no more than
> shadows flitting through the darkness outside one of the

nation's nuclear reactors. Beyond the fencing, black-clad snipers would take aim at sentries atop guard towers ringing the site. The guards tend to doubt they would be safe in their bullet-resistant enclosures; they call such perches iron coffins, which is what they could become if the terrorists used deadly but easily obtainable .50 caliber sniper rifles.

The saboteurs would break through fences using bolt cutters or Bangalore torpedoes, pipe-shaped explosives developed by the British in India nearly a century ago. The terrorists would blast through outer walls using platter charges, directed explosives developed during World War II, giving them access to the heart of the plant. They would use gun-mounted lasers and infra-red devices to blind the plant's cameras, and electronic jammers to paralyze communications among its defenders. They would probably be armed with hand-drawn maps, drawings of control panels, weak spots in the site's defenses—provided by a covert comrade working inside the plant.[30]

Once inside the plant with access to the control room they would and could easily flip a few well-learned switches, shutting pumps and operating key valves to cause a deadly loss of coolant. As the nuclear engineer David Lochbaum says, it may sound far-fetched, but "it's irreversible once that last switch is flipped."[31]

Many of the scenarios above were taken from a DOE training video for guards at nuclear power plants. As Paul Blanch, a nuclear safety expert, writes, "A knowledgeable terrorist inside a control room can cause a meltdown in fairly short order."[32]

The Nuclear Regulatory Commission, in its Design Basis Threat (DBT)—a scenario projecting the maximum threat that nuclear plant security systems are required to protect against—has always insisted that nuclear plants need only be protected against an attack by a maximum of three people outside with the help of one insider. They also have assumed that the attackers would act as a single team and be armed only with hand-held automatic rifles. Now, after 9/11, the NRC requires that guards can protect against up to eight attackers. Yet nineteen highly organized men made the attack on 9/11.[33]

The security guards at nuclear power plants complain of low morale, inadequate training, exhaustion from excessive overtime, and poor pay. They often are expected to work seventy-two hours a week, and not infrequently they go to sleep on the job.[34] They state that they would not be prepared to die to save the reactor, considering their poor compensation and the treatment they routinely receive from management.[35]

The NRC defends the poor state of security at nuclear reactors by saying that a force as large as the 9/11 team constitutes an enemy of the state, rendering the protection of nuclear power plants the job of the Pentagon and the federal government (who would never get to the reactor in time).[36]

Wackenhut Corporation, the huge security firm contracted to guard half the country's reactors, is the same company that has been contracted to test the security at the reactors. Since, by law, each plant must be tested once every three years, Wackenhut must conduct simulated test attacks an average of twice a month. In 2003, Wackenhut "attackers" tipped off Wackenhut guards about the details of the drill.[37] Wackenhut employee Kathy Davidson at Pilgrim Nuclear Station in Massachusetts was fired from her job because she complained that security was inadequate at the plants. Davidson later told *Time* magazine that, of the twenty-nine classroom exercises Wackenhut conducted to prove that guards could defend against terrorists, the attackers won twenty-eight.[38]

According to Edwin Lyman, a physicist with the Union of Concerned Scientists, a terrorist-induced meltdown could kill more than half-a-million people. Yet Marvin Fertel from the Nuclear

Energy Institute, the industry's research institute, continues to insist that only about one hundred people would be killed in such an attack, and the chances of terrorists achieving this goal are "so incredibly low it is not credible."[39]

Congressman Christopher Shays, who chairs the House Reform Committee's panel on national security and emerging threats, believes that the NRC's DBT prediction is artificially low because of economic pressures, amounting to how much security we can afford, not how much we need. Some nuclear security officials call it "the funding basis threat."[40]

A recent study by the National Academy of Sciences on the dangers posed by terrorists to 43,600 tons of spent fuel stored at the sixty-four power plant sites across the United States concluded that an additional study of security at the nation's nuclear plants is urgently needed.[41]

MELTDOWNS

A Meltdown at a Reactor Near the New York Metropolitan Region

What would a catastrophe at a nuclear power plant in the U.S. look like?

Let's consider the two large Indian Point reactors located in the town of Buchanan in Westchester County, thirty-five miles from midtown Manhattan.[42] Indian Point 2 is a 971-megawatt reactor and Indian Point 3 is a 984-megawatt reactor; the licensed operator for both plants is Entergy Nuclear. Both reactors are aging and adjacent to a very large population base: More than 305,000 people live within a ten-mile radius of the plants, and 17 million live within fifty miles. They are in close proximity to a reservoir system that waters 9 million people and to the financial capital of the world.

Apart from natural disaster, an Indian Point meltdown caused by a small group of people intent on wreaking disaster could readily be achieved in one of several ways. Terrorists with suicidal tendencies could easily disrupt the external electricity supply of the reactors, or obtain one small speed boat, pack it with Timothy McVeigh fertilizer explosives, and drive it full tilt into the two adjacent intake pipes that suck almost two million gallons of Hudson River cooling water per minute into the reactors. The plant could be shut down immediately, but this would not help because of the intensity of the heat already in the reactor. Within several hours the meltdowns would be in full swing. (Several years ago, I was in a boat, owned by the antinuclear group Riverkeeper, on the Hudson opposite the huge intake pipes of the two Indian Point reactors. Although the Coast Guard was supposed to be protecting them from terrorist intrusion, there was no sign of a Coast Guard boat during the two early afternoon hours we were within view of the pipes.)

Alternatively, a terrorist could drive a truck packed with similar explosives into a strategic area of the plant, triggering a critical situation. Concrete barriers have been erected at several nuclear power plants, but not many, and, as stated in the previous chapter, an inadequate number of guards are protecting against terrorist intrusion. A paper written by the Oak Ridge National Laboratory and the Defense Threat Reduction Agency, published in a 2004 technical journal and available on the Internet, indicates that truck bombs of various sizes would have 100% probability of success.[43]

Or yet again, after a few basic flying lessons, a novice pilot could commandeer a large passenger plane loaded with fuel and fly it into the reactor itself, destroying strategic safety systems and/or emptying the reactor of its cooling water. Or a patient individual bent on destruction could sign up for training as a nuclear power plant operator, obtain a job at Indian Point, and at a certain strategic moment, press the wrong switches and valves, removing the cooling water and initiating a meltdown from the inside.

The meltdown would follow several specific stages:

First, as the cooling water leaks or dissipates from the reactor core, the intensely hot radioactive pellets in the fuel rods overheat and swell, and the zirconium cladding oxidizes and ruptures. Volatile radioactive isotopes that have collected in the gap beneath the zirconium cladding of the fuel rods are then released as gases into the atmosphere of the containment building. These elements include the noble gases argon, krypton, and xenon, as well as radioactive iodine and cesium. This whole period, called the "gap release" phase, will last about thirty seconds.

As the core continues to heat, the radioactive ceramic fuel pellets themselves will begin to melt, and, as they do, large quantities of radioactive isotopes will be released into the reactor vessel and into the containment building atmosphere. The molten fuel will drop to the bottom of the reactor vessel; it will melt the steel and drop through the vessel onto the bottom of the containment building. This period up to when the reactor vessel is breached is called the "early in-vessel" phase and will last about one hour. At this point, the meltdown is irreversible.

When the molten fuel hits the bottom of the containment building, it will violently react with water that has collected there and with the concrete of the floor, releasing even more isotopes. This is the "ex-vessel" phase, which lasts several hours. When the molten core hits the bottom of the reactor vessel, it may trigger massive steam explosions that could blow the reactor vessel apart, simultaneously creating high velocity "missiles" that could rupture the containment building and violently expel the radioactive gases and aerosols into the outside atmosphere. If the reactor vessel is not itself blown apart, and the molten fuel just drops through the bottom of the reactor vessel onto the floor of the containment building, it could initiate a hydrogen or steam explosion, which would then rupture the containment vessel, causing a rapid purge of the radionuclide content of the containment building atmosphere into the environment.[44] This sequence involving the molten fuel penetrating the reactor vessel and the containment building is similar to what the nuclear industry calls the "Melt-through to China Syndrome," a phrase picked up as the title for the film *The China Syndrome*, starring Jane Fonda and Jack Lemmon, which was released just before the Three Mile Island meltdown.

The eventual distribution of radioactive elements is dependent upon several factors. If the zirconium fuel cladding oxidizes and burns, generating enormous quantities of heat, the radioactive plume could be elevated in the atmosphere to great heights and spread over a wide geographical area.[45] If there is no zirconium burn, the fallout will be more intense because it will be distributed over a smaller surface area. The medical and ecological consequences of the meltdown will also be affected by the prevailing meteorological conditions, including wind speed and direction, atmospheric stability, inversion systems, and whether or not it is raining, because rain efficiently precipitates radioactive fallout.[46]

In the event of a meltdown, both the Nuclear Regulatory Commission and the Environmental Protection Agency require an evacuation plan for all people living within a ten-mile radius of the reactor (In an actual event, the NRC says that it would probably only evacuate a segment of that area. The EPA has no authority to order evacuations at all.) The plan is for an off-site alarm to go off thirty minutes after the event begins, which would allow time for the operators to determine the extent of the damage, timing would coincide with the initiation of the core melt. That would leave seventy-eight minutes from the alarm sounding to the beginning of the radioactive release.[47]

The 2003 census data estimated that 267,099 people reside within this ten-mile radius, although the transient population could increase that number by 25%.[48] For those with nefarious motives, the best time for an attack would be the evening, under cover of darkness, and when the prevailing winds blow toward New York City.[49]

Prospective radiation doses are calculated in several ways. During the first week, doses of radiation will be incurred from:

- inhalation and direct exposure to the radioactive cloud;
- exposure to the radioactive particles that deposit on the ground, which emit large amounts of gamma radiation. This is called "groundshine."

Beyond the first week, doses are calculated from:

- groundshine;
- inhalation of resuspended particles from the ground;
- consumption of contaminated food and water.[50]

The calculations are truly frightening, because people in the evacuation zone will receive enormously high doses of radiation, ranging from a mean of 198 rems to a possible peak of 1,490 rems. In midtown Manhattan, doses will range from a mean of 30 to a peak of 307 rems. The radiation dose at which 50% of the population will be expected to die ranges from 250–380 rems.[51]

High levels of radiation kill the actively dividing cells in the body—hair-, gut-, and blood-forming elements. Consequently, the symptoms that will be experienced by people in Westchester County and Manhattan include: acute loss of hair, severe nausea, vomiting and diarrhea, bleeding from every orifice—nose, mouth, gums, stomach, and bowel—and massive, overwhelming infection. This collection of symptoms was first experienced by Hiroshima victims and is called acute radiation sickness.

The number of early fatalities from this syndrome within the ten-mile evacuation zone will range from 2,440 to 11,500. If it is raining and the weather conditions maximize the fallout, peak fatalities could reach 26,200 in the ten-mile zone and 43,700 in the fifty-mile zone—this latter number including people in Manhattan.[52]

Late cancer deaths, which will occur two to sixty years later, range from 9,200 to 89,500 people within the ten-mile zone, and 28,100 to 518,000 people in the fifty-mile zone—more than half a million people.[53] The early and late-stage fatalities can be reduced within the ten-mile zone if people take shelter from the radiation exposure in their houses, kindergartens, schools, and workplaces during the acute phases of the radioactive fallout. During the early stages of meltdown, isotopes with very short half-lives emit huge doses of radiation, and it is better for people to stay indoors. As the isotopes decay over several days, so the exposure relatively decreases.[54]

Evacuation tends to increase the doses received, because people are in non-airtight vehicles or on foot, inhaling the radioactive air and being exposed to the groundshine.[55] It would seem prudent, therefore, for people in the ten-mile zone to shelter in their homes, offices, and schools during the acute radioactive phase. However, as stated above, this policy is not the one advocated by the NRC or the EPA, both of which require immediate evacuation.[56] Thus, while the Indian Point nuclear power plants I and II operate at full tilt—in a country that insists on car seats and safety belts, no smoking, no swimming without a lifeguard, fire extinguishers and oxygen masks, life vests and air bags—citizens lack the most basic information about how best to protect themselves and their children in the event of a nuclear meltdown. Nor is there any official requirement to supply this information to the general population.[57]

Now imagine this scene. Over 300,000 people are running and driving away from the stricken reactor along winding Westchester roads, stuck at traffic lights and in traffic jams; all are in a state of panic, anxiety, and acute disarray, trying to reach their children at schools, their spouses and mates. Then they begin to taste a strange, metallic flavor in their mouths. They infer that each breath exposes them to deadly radioactive gases, the radio blasts out dire warnings, yet nobody knows what they are doing and nobody is in control.

And what about Manhattan? Millions of people trapped as the bridges and the Midtown, Lincoln, and Holland tunnels are totally blocked, hiding in their apartments, hardly daring to breathe.

If everyone took inert potassium iodide tablets as soon as they heard about the meltdown on the radio or TV, peak doses to their thyroids of radioactive iodine—one of the many isotopes in the deadly radioactive plume—could be reduced by 30%. It should therefore be mandatory that every person living within fifty miles of the Indian Point reactors stock potassium iodide in their medicine cabinets available to be taken at a moment's notice.[58]

Yet the NRC requires that potassium iodide tablets be available only for the ten-mile-radius population,[59] stipulating that "recommended consideration of potassium iodide distribution out to ten miles was adequate for protection of the public health and safety."[60] Compared to adults, children receive three times the thyroid dose from radioactive iodine because they have a higher respiratory rate and their thyroid is relatively small.[61] Whereas adult thyroid doses in midtown Manhattan would range from 164 rems to a peak dose of 1,270 rems, the doses for five-year-old children range from 530 rems to an unbelievable 4,240 rems. At these extremely high doses, the thyroid tissue is simply destroyed, so these children would have to take thyroid replacement hormones for the rest of their lives.[62] At lower doses that don't cause ablation of the thyroid, the risk of childhood thyroid cancer in New York would correspond to the incidence in Belarus following Chernobyl. It is clear that the mandatory stocking of potassium iodide should occur throughout Westchester County and Manhattan. However, as previously stated it is important to note that the timely use of potassium iodide will only reduce the thyroid dose by 30% so the cancer risk to children and adults will still be elevated.

And radioactive iodine is just one of many deadly, long-lived isotopes in the radioactive plume, all of which migrate to different and specific organs of the body. Food grown in the area will remain radioactive for tens to hundreds of years.

The economic consequences of a meltdown at Indian Point would be stupendous. The financial capital of the world could be rendered virtually uninhabitable, with a possible $1.17 trillion to $2.12 trillion dollars in damages accruing from attempts at decontamination, the permanent condemnation of irretrievably radioactive property, and a simple lump sum compensatory payment to people forced to relocate temporarily or permanently as a result of the meltdown. Up to 11.1 million people might have to be permanently relocated.[63] These financial estimates do not include the extraordinary economic consequences if the world's financial capital were closed forever.[64]

This section describes the effects of a meltdown at only one Indian Point reactor, but the two reactors are closely aligned, and if the structural integrity of the second reactor were damaged in the attack or the external electricity supply of both reactors were disrupted, compounded by a failure of the backup power supplies, there could be a meltdown of not one but two reactors, which would compound the tragedy many times. Then of course if the three cooling pools were also ruptured, we would face a tragedy unprecedented in the history of the human race.[65]

A DISASTER AT A SPENT FUEL POOL

Spent fuel pools, euphemistically called "swimming pools" by the industry, are typically constructed next to reactors to store massive amounts of high level radioactive waste. The United States has 103 commercial nuclear reactors, located at sixty-five sites, based in thirty-one states. Thirty-four of these are boiling water reactors (BWR) and sixty-nine are pressurized water reactors (PWR). An additional fourteen commercial reactors have been shut down and are being decommissioned. In total, there are sixty-five PWR pools and thirty-four BWR pools. Some reactors that are located at the same site, such as the Indian Point reactors, share "swimming pools."[66]

Every year at Indian Point, for example, about 30 tons of intensely thermally and radioactively hot fuel is removed from the reactor core because it is so contaminated with fission products that it is no longer efficient. In general, these pools were originally designed to hold only moderate amounts of nuclear waste because it was assumed in the original days of reactor design that this material would be transferred to a reprocessing plant, the fuel rods would be chopped up into small segments, then they would be dissolved in concentrated nitric acid, from which plutonium and uranium would be extracted to be recycled in the manufacture of more electricity.

However, President Carter put a stop to this notion in 1977 because he realized that reprocessing by the United States would encourage other countries to do the same, and extraction of plutonium from radioactive fuel sets a dangerous precedent because it is the fuel for atomic and hydrogen bombs. In response, the Congress passed the Nuclear Waste Policy Act in 1982, pledging that the federal government would provide, at its expense, a deep underground storage facility for commercial spent fuel. Transfer of the fuel to this repository was scheduled to commence in 1998.

The Department of Energy then identified a so-called suitable mountain in Nevada called Yucca Mountain, which has subsequently been drilled and excavated to receive the waste. Apart from the fact that the people of Nevada are adamantly opposed to the notion of a radioactive Yucca Mountain, so many technical and scientific problems have plagued the project that the site may never be ready to accept the waste. The most recent estimate for a start date is 2015. (Problems with Yucca Mountain are addressed in the next chapter.)

As the reactors continue manufacturing ever more nuclear waste, and the swimming pools become overloaded, the NRC has licensed reactor owners to re-rack and over-pack the fuel, approaching densities similar to those in the reactor itself. This is potentially a very dangerous situation, because there is a chance that the densely packed fuel could reach critical mass, triggering a meltdown in the cooling pool. In order to prevent a criticality, this "dense-packed" fuel is now maintained in a subcritical state by enclosing each fuel assembly in a metal box made of the neutron–absorbing material boron. But in the event of a loss-of-coolant accident, convective air-cooling, which is very effective in open-spaced pools, would be ineffective in a dense-packed pool. Even without going critical, the spent fuel can overheat and melt if the cooling water is lost rapidly enough. Indian Point is about to begin implementing dry cask storage. Its cooling pools have been re-racked and re-racked, and are overloaded.[67]

These spent fuel pools house enormous amounts of radiation. There is almost twice as much cesium 137 in a ton of spent fuel as in a ton of reactor fuel. (Reactor cores each contain about 5 million curies of cesium in 80 tons of uranium, whereas spent fuel pools contain 35 million curies of cesium incorporated in 400 tons of spent fuel.) A meltdown in a spent fuel pool could be catastrophic—much worse than a meltdown at a nuclear reactor. Of all the radioactive elements in a spent fuel pool, cesium 137 is the most worrisome because it accounts for 50% of the radioactive inventory in fuel that is ten years old.[68] As we know, cesium is a volatile isotope. Readily dispersible, with a thirty-year half-life, it is radioactive for 600 years. But other nasty isotopes would be released as well, including plutonium and all its alpha-emitting relatives, plus cerium, technetium, tritium, strontium 90, and many others. The studies below of casualties caused by a pool melt refer to the medical effects of cesium 137 only, so they are actually gross underestimates over the long term.

The NRC performed a study in 1997, which calculated that a fire at a spent fuel pool could produce between 54,000 to 143,000 cancer deaths and would render 2,000 to 70,000 square kilometres of agricultural land uninhabitable. In addition, $117 billion to $566 billion would need to be spent evacuating hundreds of thousands of people from contaminated areas. The study, by Alveraz and others, determined that if just 10% of the cooling pool cesium 137 were released by fire, the area contaminated would be five to nine times larger than the area affected to a similar degree by Chernobyl. If 100% were released, the contamination would affect an area about seventy times larger than that of Chernobyl.

Amazingly, many spent fuel pools at boiling water reactors are built atop the reactor building or above ground level, making them very vulnerable to plane crashes and other terrorist attacks. A turbine shaft of a high-speed jet fighter or a large passenger jet could easily penetrate the wall of the cooling pool, destabilize the pool supports, or even overturn the pool, allowing the cooling water to drain away. A fuel-air explosion from aerosolized jet fuel could create a fireball that would collapse the building above the pool, destroying the pool. A very hot fireball alone could also evaporate some of the cooling pool water.

Other unpredictable events hover menacingly over these radioactive Pandora's boxes. Earthquakes threaten to disrupt the fuel assembly geometry, and one of the spent fuel casks that are routinely passed across the top of the pool could accidentally be dropped, severely damaging the pool. These pools are also extremely vulnerable to being punctured by a shaped-charge antitank missile.

Once the water in the pool drops below the top of the fuel, the gamma radiation would be so intense—10,000 rems at the edge of the pool and hundreds of rems in certain other areas—that lethal doses would be incurred in less than one hour. This would prevent any efforts by staff to intervene or to contain the situation. Amazingly, the NRC has virtually ignored the dangers that these cooling pools present. It does not require reactor operators to prepare for any of these emergencies, either with redundant safety systems or emergency backup water-cooling systems.

The cooling pool problem is further potentiated because many utilities now disgorge all their fuel into the cooling pool every twelve to eighteen months in order to inspect the reactor and its parts in a short space of time. This new operation is expedient and is performed in the name of "efficiency" or saving money. But it makes the situation at the cooling pools more tenuous because the fuel is intensely hot, and a loss-of-coolant accident at this time would be catastrophic. Before this new "efficiency" model arrived on the scene, only 30% of the highly radioactive spent fuel from the reactor core was ever placed into the cooling pool at a given time.

If loss of coolant water occurs at a cooling pool, the boron boxes that house the densely packed spent fuel block the free circulation of air among the fuel rods. In such an event, a freshly discharged reactor core in the pool would generate so much heat within one hour that the zircaloy cladding would rupture as the fuel elements expanded. When it reached 900 degrees centigrade, the zircaloy would burst into flame.

Because these events are so potentially catastrophic in nature, it is imperative that the Congress, the nuclear industry, and the NRC decide immediately to remedy the situation. Alveraz and his colleagues have arrived at a temporary solution that would mitigate an enormous radioactive release from a spent fuel pool in the event of an accident or terrorist attack. Their plan directs that fuel five years old or older be removed from the pool and placed in dry cask storage at the reactor site, removing the risk of an attack or accident to a huge collection of volatile radioactivity in a single site. A transfer of all spent fuel older than five years would reduce the cesium inventory in the pool to a quarter its current size, or two (rather than eight) times that in the reactor core. On average, thirty-five casks would be necessary at each reactor.

Currently, thirty-three reactors have resorted to dry cask storage and twenty-one are in the process of setting this up. This would then allow the utilities to return to open-rack storage of their fuel rods, thus allowing air to circulate freely between the fuel rods, affording some degree of safety and cooling factor in the event of an accident or attack. The dry casks, which are passively cooled by the natural convection of air powered by the innate heat of the rods, are stored on concrete pads in the open. As such, they too are vulnerable to terrorist attack, but the release of cesium from a single cask would be very small compared with a fuel pool melt. However, should a powerful bomb explode over a fleet of dry casks, many of them could be ruptured.

At the moment 200 casks a year are being fabricated, which could store 2,000 tons of spent fuel. Spent fuel that is less than five years of age is unsuitable to be stored in dry casks because it is still so hot that it could spontaneously melt. If all but the last five years of the discharges were to be dry stored, 300 dry casks per year would be needed to unload 35,000 tons of high-level waste over the next ten years. If this does not take place, the total radioactive inventory in the fuel pools is predicted to be a huge 60,000 tons by 2010, of which about 45,000 tons will be in the dense-packed cooling pools. The danger with densely-packed pools is that if they lose their cooling water, the fuel could catch fire, whereas the fuel in dry casks is less concentrated and therefore less likely to burn.

Nuclear power plant owners will obviously not want to cover the extra expense of dry cask storage because in a deregulated market they will be unable to pass this cost onto the consumers without fear of being undersold by competing fossil-fuelled plants. In order to prevent delay in the implementation of these guidelines, the federal government should offer to underwrite the expenses for these storage casks and to pay for the necessary security upgrades. Once again, the nuclear industry will be leaning upon the federal government to cover its expenses.

Such disaster scenarios are not limited to the United States, of course. In England soon after the 9/11 attacks in New York, Greenpeace commissioned a series of three reports that examined the results of an aerial terrorist attack on the nuclear complex at Sellafield. (The Sellafield complex comprises nuclear reactors, re-processing plants, and high-level waste storage tanks containing 1550 cubic meters of liquid waste plus tens of tons of separated plutonium.) To their horror, they discovered that three-and-a-half-million people could be killed. Greenpeace was so shocked by these results that they sat on the data for over a year, unsure what to do about them, before they decided to release it. Dr. Frank Barnaby, a former scientist at the British nuclear establishment at Aldermaston, concluded that an attack by a jumbo jet onto the Sellafield plant could cause a radioactive fireball over a mile high. It would only take four minutes for a plane to be diverted from its regular flight path to the Sellafield nuclear complex in Cumbria and, in the event of an attack, twenty-five times as much radiation as that emitted from Chernobyl would likely be released.[69]

5

Yucca Mountain and the Nuclear Waste Disaster

Never in its sixty-five-year history has the nuclear industry taken responsibility for the massive amounts of profoundly lethal radioactive waste that it has continued to produce at an ever-increasing pace. As it is impossible for mere mortals to fathom the concept of infinity of time and space, so it is impossible to comprehend the true gravity of mutagenic carcinogens lasting for half-a-million years.

Radioactive waste comes in various forms and guises, classified according to concentration of isotopes, type of isotopes, and origins of the material. Of these, the most dangerous is high-level waste, extremely concentrated waste, pulsing with intensely energetic radioactive elements emanating both from the production of plutonium for nuclear weapons (91 million gallons in total) and from radioactive spent fuel from nuclear power reactors (52,000 metric tons in 2006).[1]

Initially, the Department of Energy proposed in the early 1970s to store high-level waste in salt domes in Lyons, Kansas, but it was discovered that the domes had been accidentally punctured by gas exploration holes. Other sites were explored and ruled out, largely in response to political pressures from an aroused public. In 1982, Congress passed the Nuclear Waste Policy Act, promising to take responsibility for the radioactive waste from nuclear power plants, and in 1987 Congress designated Yucca Mountain in Nevada as the primary repository.

Initially, it was thought that the Yucca site would be simple and that the geology and site characterization would be easy. Yet in retrospect, Yucca Mountain, a volcanic remnant composed of pumice or layers of volcanic ash, with a complex geology, was a particularly poor choice. To this day, Yucca Mountain has yet to receive a single shipment of nuclear waste.[2]

The stated requirement of a geological storage site is to prevent the leakage and seepage of waste for at least 500,000 years. The EPA standards require storage to the time of the peak dose. This may be as many as 500,000 years in the future.[3] That mandate will never be achieved at the Yucca Mountain site for several reasons:

- Contaminated water from corroded casks could seep into the groundwater and spread to Amargosa Valley, which may communicate via a hydrological pathway to Death Valley, and spread to nearby farming areas and protected species habitat, and produce radioactive springs.[4]
- A volcanic event may lead to magma intrusion into the tunnels where the waste is stored, melting the cannisters. If the volcanic event opened a path to the surface, radioactivity could be spread around the landscape.
- Radioactive chlorine 36 has been found deep inside the so-called waterproof mountain. This isotope could have come only from atmospheric nuclear testing performed in the 1950s and 1960s, meaning that water penetrated the mountain in fewer than fifty years, thousands of times faster than was estimated by the DOE.[5]
- Much more water was discovered inside the mountain than originally estimated—conditions were supposed to be relatively dry so the metal casks containing the radiation would not corrode over time. It is expected that the climate will

become much wetter in the future. This will likely increase infiltration.[6]

• Yucca Mountain is located in an active earthquake zone, and at least thirty-three known active faults pass within twenty miles of the Mountain, some within the proposed repository itself. In June 1992, an earthquake measuring 7.4 on the Richter scale hit Yucca Valley in Southern California.[7] Two days later, an earthquake measuring 5.2 caused $1 million of damage to the DOE building located six miles from Yucca.

• Part of Yucca Mountain is below the Nellis Air Force Base flight training region. This is where new military jets and pilots are tested. It is also used for war exercises involving both US pilots and pilots from other countries. Crashes are not unusual.[8]

Some scientists have compared the Yucca Mountain Program to "NASA before the *Challenger*."[9]

In such an inadequately chosen mountain, huge quantities of radioactive waste will need to be stored for thousands of years. Temperatures inside the repository will be above boiling point for 1,250 years, with temperatures reaching 662 degrees Fahrenheit inside the canisters and 527 degrees or far above boiling point in the rock holding the canister. One spent fuel assembly contains ten times the amount of long-lived radiation released by the Hiroshima bomb; the mountain is to house 140,000 assemblies.[10] However this data is subject to revision as the whole project is currently (January 2006) under revision.

The following information is gleaned from the Nevada State web site titled "Chronology of Selected Yucca Mountain Emails." The DOE, under the 1982 Nuclear Waste Policy Act, was obliged to report to Congress the bad news that Yucca Mountain's geology is inappropriate to contain the nuclear waste. But by the time these derogatory assessments had been made, the DOE, commercial

nuclear industry, and many members of Congress were clearly intent on just making the Yucca Mountain repository work no matter what.[11] Rather than end this venture for the DOE Yucca
Mountain bureaucracy and its contractor, the project had taken on
a life of its own. The DOE shifted from assessing geological storage to designing a man-made waste package that would prevent
water from reacting with the radioactive waste—even though it is
a golden rule that radioactive waste storage facilities require "defense in depth"—that is, when the containers fail and radiation es-
capes, the geological environment must prevent further radioactive
escape. DOE engineers developed a bizarre scheme, consisting of
a titanium "drip shield" to be placed by remote control over the
canisters some hundred years hence by our descendents, just before the repository is finally closed—when the tunnels themselves
may well have collapsed. The DOE then decided to ignore the
troublesome site selection geological guidelines.[12]

As the DOE decided to rely on the waste package and drip
shield instead of geological integrity, the problems with corrosion
became crucial to the exercise. But the DOE does not have long-
term corrosion data on C22, the nickel-based alloy used for the
containment canisters, as it has been in existence for only a few decades. It is impossible to extrapolate from a few decades to tens of
thousands of years when the fundamental corrosion mechanisms
are not well understood. Nevertheless, glossing over the science, in
typical fashion, and relying on inadequately supported models, the
DOE publicly reassured everyone that the integrity of these materials would be maintained for tens of thousands of years.[13]

Meanwhile, no matter how the DOE manipulated its computer codes, because the site was so poor, the peak dose to the
public once the canisters failed was too high to meet the EPA
guidelines. So late in the time frame that these decisions were being made, the NRC and the EPA decided to change the guidelines
limiting the period of regulatory compliance to 10,000 years

(down from hundreds of thousands of years), which then allowed the DOE to qualify for a license on the basis of its flawed corrosion-resistant packages.[14]

In July 2004, however, the federal court of appeals (DC Circuit), ruled that the period of regulatory compliance had to extend to the peak dose, when the canisters would lose their integrity and start leaking, whenever that was, even after 10,000 years. The court told the EPA and the NRC that the DOE could not rely on the waste package alone.[15] The EPA under George W. Bush then issued new regulations that elevated the allowable dose so that the DOE would pass the necessary qualifications. Bizarrely, they decided on a two-tiered regulation that would limit exposure to 15 millirems per year per individual near the proposed Yucca Mountain facility for 10,000 years, then would suddenly allow an *increase* of the dose to 350 rems per year—three and a half times the dose presently allowable to members of the public from all sources by the NRC—for up to 1 million years. (This is 23 times higher than the 15 millirem standard that the EPA previously set as a safe limit.) (A standard chest X-ray is 10 millrems.) Senator Harry Reid from Nevada labelled this policy "voodoo science and arbitrary numbers."[16] And Paul Craig wrote in his paper on Yucca Mountain, "Rather than starting with an idea as to what is safe enough, EPA is asking what kind of standard will let Yucca proceed."[17] When asked if this standard is relevant or could be maintained so far in the future, the EPA's Jeffrey Holmstead said, "That's a pretty darn good question. . . . We do the best job given all the science we have."[18]

Meanwhile, the DOE commissioned the U.S. Geological Service to provide the answers on water infiltration in the mountain. In practice, though, the Geological Services engineers were reporting to private contractors, and these engineers were under heavy pressure to produce results. In March 2005, after a DOE contractor came upon incriminating e-mails, the DOE announced that certain USGS employees might have falsified data on Yucca Mountain.

The e-mails were turned over to Congress because of a demand by Congressman Jon Porter, who chaired a hearing on this issue. At the very least, it is clear that these scientists were not following established quality assurance (QA) procedures to assure the validity of their work. Nevertheless, the secretary of energy supported the selection of Yucca Mountain as the nation's high-level waste repository site, repeatedly characterizing the engineers' reports as "good science."[19] Following are selected extracts (reproduced as written, without corrections) from these e-mails.

> I am not suggesting the Cl-36 data doom this site. . . . The radio-toxicity of spent fuel is so high that even if a very small amount of water can contact the waste and reach the accessible environment, there will result a substantial dose.
>
> The Cl-36 data suggests there are fast paths from the surface to at least the repository horizon. . . . If the waste packages last only a few thousand years, the calculated dose rate would exceed 100 mrems per year in less than 10,000 years.
>
> His analyses suggest that the precipitation at YM will increase as the result of global warming.
>
> I think that arguing that drip shields don't help you either (you will get some water contacting the waste . . . that means that the eventual dose is 1 rem/year—or 10 rem/year. . . . I think it is impossible to show the doses will be less than 10 rem/year even for drip shield. . . .).
>
> These guys are going to assign probability distributions that keep the expected values in the right place. . . . There is no generally accepted way to calculate the behaviour of the geohydrological environment as a function of heat releases from the disposed site. The time constant of the processes are such that there can be no confirmation of any calculation process for many years. . . . Observations at several

DOE sites confirm that radionuclides tend to move further and faster than model predictions.

The issue is the inertia of a human organization [the DOE] that is large, old, isolated, behind the times, in charge of its own money, self grading and self satisfied. . . . We know everything we are going to know about the site, and it's enough. All that's left is 1. how to install a drip shield and gravel and 2. how to make long-life waste packages.

Wait till they figure out that nothing I've provided them is QA. If they really want the stuff, they'll have to pay us to do it right.

What if you just download the raw files from [blacked out] and say you used those? Do they need to know more than that? You don't really need to do an analysis just say this is the data I used. Maybe that would work.

I've deleted the lines from the "official" QA version of the files (which do have headers). In the end I keep track of 2 sets of files, the ones that will keep QA happy and the ones we actually used.

There is of course no scientific notebook for this work. All work is in the form of electronic files. . . . I can start making something up but then the [blacked out] projects will need to go on hold.

I don't have a clue when these programs were installed. So I've made up the dates and the names (see red edits below). This is as good as its going to get. If they need more proof I will be happy to make up more stuff.

What good is QA if there is no data or analysis to QA? Do we just QA the QA?[20]

Another important aspect of waste management not covered in depth here, but critical if Yucca Mountain ever becomes operational, is the transportation of radioactive waste along the highways and railways of the United States to Yucca Mountain. It is estimated that it would take thirty years to deliver 70,000 metric tons of civilian spent fuel, that Yucca Mountain is authorized to receive— 63,000 metric tons accumulated from civilian reactors and 7,000 metric tons from military waste. However, the quantity of waste to be produced from the present generation of U.S. reactors will far exceed 63,000 metric tons, let alone if new reactors are built.[21] And it is predicted that there could be fifty accidents a year, three of them serious, with radioactive releases. There is no prohibition of radioactive shipments through highly populated areas, nor during serious weather conditions. All eleven of the casks currently used by the DOE for radioactive transport have been found to be defective.[22]

Yet as recently as August, 2005, Craig Stevens, a DOE spokesman, confirmed the Bush administration's commitment to proceeding with the Yucca Mountain Project; it plans to submit a formal application for a licence to the NRC.[23]

6

Generation IV Nuclear Reactors

The nuclear industry classifies its reactors according to "generations."

GENERATION I REACTORS

Primitive by today's standards, Generation I reactors were developed in the 1950s and 1960s and are fueled with natural, unenriched uranium. Only eight Generation I reactors, all in the United Kingdom, remain in operation today.[1]

GENERATION II REACTORS

The majority of the four hundred and forty-one nuclear power reactors currently operating in a total of thirty-one countries are designated Generation II reactors. Generation II reactors come in different varieties. The first is the classic light water reactor, which comes in two forms: The Pressurized Water Reactor (PWR) and the Boiling Water Reactor (BWR), both of which are in operation in different parts of the United States. Both light water reactors are cooled by ordinary water.[2]

The pressurized water versions are afflicted with several problems. Originally designed to propel nuclear submarines, PWRs operate at higher temperatures and pressures than other reactor designs. These conditions accelerate corrosion of many vital components including the steam generators. They can also initiate cracking at the site of the reactor head holes, through which the control rods operate. The most serious of these problems manifested in 2002 at the Davis-Besse plant in Ohio, when the reactor pressure vessel cracked and serious corrosion took place, bringing the reactor perilously close to a meltdown.[3]

The BWR design is also beset with significant corrosion and cracking events, which have become particularly severe in BWR reactors in Germany. In addition, the plumbing in boiling water reactors is substantially more complex than that in the pressurized water reactors, leading to other kinds of structural and functional problems.

Another common Generation II reactor is the pressurized heavy water reactor, forty-four of which are currently operating in seven countries.[4] The most common reactor in this group is the CANDU reactor, originally designed in Canada. It is fuelled by natural uranium and cooled by heavy water. In addition to the fact that these reactors generate much larger quantities of spent fuel than light water reactors, many have encountered severe safety and economic problems and have been closed.[5] CANDU reactors generate large quantities of tritium as a by-product of heavy water irradiation and expel large quantities of tritium to the biosphere. In April 1996, a massive 50 trillion curies of tritium were released into Lake Ontario from a leak at a heat exchanger at the Pickering-4 station. Lake Ontario is a very large body of water, and the tritium would have been rapidly diluted. However, many people get their drinking water from the lake, and if they lived near the outflow from Pickering they could have ingested tritium. This isotope also bio-concentrates in the food chain, so people who catch and eat fish from the lake could also ingest tritium. As the half-life of tritium is 12.4 years, it will remain in the lake for more than one hundred years.

CANDU reactors also generate large quantities of the isotope plutonium 239, more desirable for the manufacture of nuclear weapons, making the production of nuclear weapons—from this by-product of nuclear energy—relatively easy for host countries. India manufactured nuclear weapons from its CANDU spent fuel in the 1990s.[6]

Russia designed its own reactors—which are plagued by many intrinsic safety flaws. Most egregiously, the graphite used to slow down neutrons in the reactor core can ignite during an accident and burn at extremely high temperatures. When this happened during the Chernobyl accident, the situation induced a massive thermal updraft, which lifted the radioactive isotopes high in the air, to be spread around the globe as lethal fallout.[7]

Britain has specialized in two reactor designs: the MAGNOX (the name is derived from the magnesium alloy casing surrounding the fuel rod) air cooled graphite-moderated, natural uranium reactor, and the Advanced Gas Cooled Reactor (AGR).[8]

Neither of these designs includes secondary containment vessels, so both have a potential for large radioactive releases. The MAGNOX reactor is regarded as particularly dangerous due to many safety deficiencies. It is currently being phased out of use.[9]

Despite the serious and potentially fatal problems experienced by many of these Generation II reactors, the nuclear industry is proceeding to postulate and propagate new reactor designs that promise to be, in many ways, even more dangerous.

GENERATION III REACTORS

Generation III reactors—or "advanced reactors"—with so-called "evolutionary" designs, are actually only modifications of the light water Generation II reactors currently operating in the United States and many other nations. A 2005 Greenpeace study on nuclear reactor hazards notes that most of the Generation III reactors appear to be a heterogeneous collection of different reactor concepts, some of which are barely evolved from the current Generation II reactors. Specific modifications are aimed solely at cost-cutting and improved economic performance, although publicly the industry fallaciously claims that these reactors are safer, hoping to improve public acceptance of nuclear power.[10]

With the recent NRC approval of the AP-1000 design, four reactor designs have been certified by the NRC in the United States, including the General Electric Advanced Boiling Water Reactor (ABWR), the Westinghouse System-80+, and the Westinghouse AP-600 and AP-1000 Pressurized Water Reactor (PWR).[11] The only three commercial Generation III reactors in operation at the moment, however, are all based in Japan, and another is under construction in Finland.[12] Whereas the first two designs differ little from current light water reactors, the AP-600 is specifically designed to reduce capital costs "by eliminating equipment which is subject to regulation." One of these cost-cutting features relies on dual purpose equipment—systems that supply feedwater to the steam generators *both* during normal operations and during an accident. This design also incorporates so-called passively safe systems that rely on gravity instead of motor driven pumps to initiate emergency operations during an accident.

Because concrete and steel account for a large part of the capital costs in a new reactor, Westinghouse has reduced the size and strength of the containment and other safety-grade features. However, despite these economic modifications, the AP-600 reactors failed to attract customers.[13] Westinghouse then designed a larger version of the AP-600, the AP-1000, a reactor that almost doubles the power output of the AP-600 while avoiding a proportionate increase in construction costs. This modification, however, seriously compromises the safety of these reactors, because it allows a dangerous amount of pressure to build within the weakened containment structure, which could rupture the vessel and cause a meltdown.[14]

Nevertheless, a newly formed consortium, NuStart Energy, is currently choosing sites for construction of two Generation III reactors in the United States—a General Electric ABWR reactor and the Westinghouse AP-1000, despite its design flaws.[15] The sites under consideration are the Grand Gulf Nuclear Station in Port Gibson, Mississippi; the River Bend Nuclear Station in St. Francisville, Lousiana; the DOE's Savannah River Site near Aiken, South Carolina; and two Constellation Energy plants: Calvert Cliffs Nuclear Power Plant in Lusby, Maryland, and Nine Mile Island Point Nuclear Station in Scriba, New York.

About twenty different designs for additional Generation III reactors are under development. Some are expected to be built and operational by 2010.[16]

GENERATION III+ REACTORS

One of the potentially more dangerous Generation III reactors on the drawing board is the Pebble Bed Modular Reactor (PBMR), sometimes referred to as Generation III+. This PBMR is another attempt to reduce capital costs by an "inherently" safe design that requires fewer safety features. These reactors were contemplated during the 1970s into the late 1980s, and prototype plants were developed, which operated for short periods of time in the United States, the United Kingdom, and Germany.[17] Then the nuclear industry became plagued by cost over-runs and two dreadful accidents.

The Pebble Bed Reactor is a high temperature, gas-cooled reactor (HTGR), which operates at 900 degrees centigrade and is cooled by helium gas circulating at high pressures. The fissionable fuel consists of billions of microspheres or kernels of enriched uranium oxycarbide, which is coated with two layers of pyrolytic carbon and one layer of silicon carbon; this combination is sealed in graphite spheres. Each reactor will contain up to 10 billion uranium fuel kernels covered by the graphite or carbon coating, which is supposed to prevent fission products escaping from the radioactive kernels.

A total of 400,000 of these tennis ball–sized graphite fuel assemblies or "pebbles" will be fed continuously from a fuel silo into the reactor core. The nuclear industry postulates that the slow circulation of the pebbles through the core will produce a small core size, which will reduce excessive reactivity in the core, while lowering the power density, thus minimizing the danger of a meltdown. These conditions, they say, will make the PBMR so safe that it will not be necessary to construct a containment building. (Conveniently, this means that the reactor will also be much cheaper to build.) It is even claimed by prospective reactor operators that this particular power plant is "walk-away-safe," meaning that operators could leave the site and the reactor would never enter a critical condition or crisis.[18]

Should the core temperatures for any reason—during an unexpected accident triggered by human or mechanical error—exceed 1,600 degrees centigrade, however, the carbon coating would fail (at the same temperature that zirconium would oxidize and burn, as in most other currently operating reactors) thus initiating the release of massive quantities of radioactive isotopes.[19] The radioactive kernels themselves would melt if temperatures went above 2,000 degrees centigrade. This situation would induce a graphite fire similar to Chernobyl.

Other problems include the design of the cooling system. If air were to enter the primary helium circuit, the carbon coating of the kernels could spontaneously ignite, causing a severe graphite fire with catastrophic radioactive releases similar to Chernobyl.[20] And although the reactors themselves will be located underground, the two steam turbine generators and the reactor cavity cooling system will be above ground, making them extremely vulnerable to sabotage and fires.[21]

Other problems beset the PBMR:

- It is difficult to prevent radioactive helium leaking from the PBMR reactor.[22]
- It is difficult to fabricate hundreds of thousands of fuel pebbles without imperfections.[23]
- The PBMR creates less low-level waste but a greater volume of high-level waste.[24]
- PBMRs achieve their economic advantages by replacing the steel-lined, reinforced-concrete containment structures with a far less robust enclosure building. Even the NRC's Advisory Committee on Reactor Safeguards calls this "a major safety tradeoff."[25]

GENERATION IV REACTORS

Whereas the Generation III and III+ reactor designs represent evolutionary changes from their Generation I and II predecessors, the Generation IV reactors are revolutionary. The Generation IV reactor designs rely on fuel and plant performance that have not been tested, yet alone proven to be achievable. For example, many of these designs require metals to resist corrosive conditions far more challenging than those experienced to date. It seems doubtful that the Generation IV designs will successfully meet these daunting challenges and will instead experience higher costs and lower safety levels than anticipated by the nuclear industry.

Nuclear Fuel Cycles

Hand in glove with the call for new nuclear power reactors has been the renewed call for "closing" the nuclear fuel cycle. The United States uses an "open" fuel cycle in which uranium is mined from the ground, enriched to the concentration needed for Generation II light water reactors, and then discarded with its waste products. Closing the fuel cycle involves "mining" the spent fuel from reactors for its plutonium, which will then be re-utilized in nuclear power reactors. This concept is called the Plutonium Economy.

Plutonium is produced as a byproduct from the generation of electricity by all nuclear power reactors. Approximately 95% of the uranium in nuclear reactor fuel is the isotope U–238. When U–238 atoms capture neutrons bouncing around the reactor core, they are converted into the plutonium Pu–239 isotope, which can then be used to fuel nuclear power reactors. It can also be used to make nuclear weapons.

The physical process for extracting and using plutonium, however, is complicated and very dangerous. First, spent fuel rods are chopped into small pieces and dissolved in a vat of concentrated nitric acid. Plutonium and unused uranium are then recovered from this radioactive cauldron. This is called reprocessing. Approximately 94% of the spent fuel from a light water reactor is composed of unused uranium, 1% is plutonium, while intensely radioactive fission products account for 5%. The plutonium and some recovered uranium are fashioned into ceramic pellets, which are then packed into fuel rods and placed into a reactor core.

However, contrary to its name, a closed fuel cycle does not eliminate the need for disposal of highly radioactive wastes. A closed fuel cycle incorporates a reprocessing plant and fuel fabrication plant at the reactor site, to reprocess the spent fuel, to extract plutonium and uranium, and to recycle these elements back into the reactor core. Reprocessing, as we have discussed, is the most dangerous part of the nuclear fuel cycle, besieged by dangerous environmental radioactive releases, worker contamination, and terrible problems implicit in the disposal of millions of gallons of intensely corrosive acidic radioactive liquid waste, which remain after the uranium and plutonium extraction. An MIT study estimated the cost of a closed fuel cycle would be 4.5 times higher than that of an open, non-reprocessing cycle.[26]

Apart from the higher expense, it is terribly dangerous to reprocess spent fuel to obtain plutonium that can then also be used by countries to make nuclear weapons.

Although plutonium is produced as a byproduct from operation of Generation I and II nuclear power reactors, the majority of these reactors are light water reactors, which are also called "thermal reactors" because they require "slow" or "thermal" neutrons. When atoms fission, or split, they release energy and fast neutrons. The uranium fuel used in Generation I and II reactors is not designed to handle fast neutrons, so water or graphite is used to slow down or "moderate" the fast neutrons. This moderation enables the "slow" neutrons to interact with uranium fuel atoms to cause additional fissions, which produces more thermal and electrical generation. But moderation impedes the efficient conversion of U-238 atoms into plutonium, since those atoms prefer fast neutrons.

Some of the Generation IV reactors are "breeders" while others are "fast reactors." Such reactors were specifically designed to use fast neutrons to promote the conversion of U-238 atoms into Pu-239 atoms. These reactors therefore do not use water or graphite to slow or moderate neutrons. In the breeders, the conversion process is designed to be self-sustaining, so that the problem of scarce supplies of uranium will be mitigated. Plutonium would be harvested from the spent fuel, placed in reactors, and fissioned producing electricity, thus solving the energy problem that confonts the human race. But this utopian dream did not eventuate, because major problems developed with breeder and fast reactors.

Breeder and Fast Reactors

Nuclear power reactors are prone to two types of accidents: (1) loss of coolant accidents, and (2) reactivity excursion accidents. Loss of coolant accidents occur when the heat produced by the reactor core cannot be removed, causing the nuclear fuel to overheat and melt. Reactivity excursion accidents occur when control of the reactor core is lost, causing a runaway nuclear reaction and the release of considerable amounts of energy. The 1957 accident at Windscale in the United Kingdom occurred because of operator error, when the reactor overheated and the graphite moderator caught fire.[27] The 1979 accident at Three Mile Island in the United States was also a loss of coolant accident, while the 1961 accident at SL-1 at the Idaho National Engineering Laboratory in the United States and the 1986 accident at Chernobyl in the Ukraine were reactivity excursion accidents. As indicated below, breeder and fast reactors can increase both the probability and the consequences from loss of coolant and reactivity excursion accidents.

Breeder reactors are designed to manufacture large quantities of plutonium. They are usually cooled with liquid sodium that removes the heat but does not slow down the fast neutrons. However liquid sodium metal burns and explodes when exposed to water or air. Great care (i.e., cost) must be taken to ensure zero leakage from vessels, piping, valves, and pumps containing liquid sodium because the resulting explosion can trigger a loss of coolant accident. In addition, it is dangerous to use fast neutrons rather than slow neutrons. Slow neutrons provide greater response time for equipment and operators, in the event of a reactivity excursion accident, which could possibly be stopped before reaching a tragic release of radiation. Fast neutrons, on the other hand, provide very little time or margin for error. And finally, the operation of a reactor core with higher concentrations of plutonium also means that the combined fission products and plutonium are a nastier brew than produced from today's reactor cores, permitting a deadlier radioactive cloud to be released during an accident at a breeder or fast reactor.

Closing the Nuclear Fuel Cycle

"Closing" the nuclear fuel cycle obviously will contribute to nuclear weapons proliferation. The total amount of plutonium required to operate the fuel cycle at a breeder reactor, its reprocessing plant, and fuel fabricating facilities is a massive 15 to 25 tons.[28] Since plutonium is the fuel of choice for nuclear weapons, these reprocessing facilities are very dangerous sources. The minimum amount of plutonium necessary to make an atomic bomb is only 1 to 3 kilograms or 2.2 to 6.6 pounds. Breeder reactors and reprocessing plants therefore are a standing invitation for the proliferation of nuclear weapons.[29]

The actual breeding of new plutonium fuel at nuclear reactors has never eventuated because of the extraordinary safety and waste management regimes required to handle plutonium and because these reactors never operated successfully. Instead, the cost of reprocessing and fabricating the fuel has increased dramatically over the past three decades, making it cheaper and safer to simply mine natural uranium than recycle plutonium.[30]

Fast reactors are similar to breeder reactors because neutrons in the reactor core are not moderated. Fast reactors typically do not contain as much U-238 in the reactor cores for conversion into Pu-239, so they do not produce as much plutonium as breeder reactors. Fast reactors are an excellent way, says the nuclear industry, to get rid of very long-lived plutonium, while at the same time generating electricity. This technique is called "actinide management." If a fast neutron reactor operates without the U-238 blanket installed in a breeder reactor core, five additional tons of plutonium can be placed in its reactor core, 10% of which will be converted to fission products.[31] This is called "waste burnup" or "transmutation" of plutonium. But only 10% of the plutonium is converted to deadly, long-lived fission products such as strontium and cesium, which last for 600 years, while 90% of the longer lived plutonium remains.

Fast neutron reactors are twice as expensive as light water reactors, which themselves are 60% more expensive than coal and gas fired plants and wind energy, and these "waste burn-up" reactors will have no impact upon the hundreds of tons of plutonium already stockpiled around the world.[32] The nuclear industry loves the concept of fast reactors because they say they can get rid of plutonium, which is a lode stone around their necks because it is a) so toxic, b) lasts for 500,000 years, and c) is fuel for atomic bombs. But as noted above only 10% of plutonium is fissioned in a fast reactor and that 10% is converted to deadly fission products—cesium and strontium, which themselves last for 600 years. The industry is pulling a fast one on an unsuspecting public.

Despite these extraordinarily expensive and dangerous plans, the nuclear industry egregiously claims that its Generation IV nuclear reactors will be the dream fuel providers: safe, proliferation resistant, economically competitive, and producing no greenhouse emissions. They even call these reactors "sustainable," a term applicable only to renewable energy sources and conservation.[33] Such claims are as baseless today as "too cheap to meter" was fifty years ago.

People with an intimate understanding of the nuclear industry are severely opposed to a nuclear renaissance. For example, David Lochbaum, a nuclear engineer who was employed by the nuclear power industry for seventeen years before he left in 1996 to become the Nuclear Safety Engineer for the Union of Concerned Scientists, recently spoke before the House Government Reform Subcommittee on Energy and Resources on the Next Generation of Nuclear Power. During that testimony he concluded that:

- It is inappropriate for the industry to talk about Generation IV reactors when neither the United States nor the rest of the world has a Generation I high-level waste disposal site, or has successfully operated even a Generation III reactor. His recommendation—the federal government must create a repository for high-level nuclear waste before it licenses the next generation power reactors.
- Urgent reform of the NRC is needed. Lochbaum cited NRC inadequacy in inspection procedures for the Davis-Besse nuclear plant in Ohio, which came so perilously close to a meltdown in 2002. In that context he said that the NRC had failed to implement at least 25% of the "high-priority" lessons it had experienced from past problems at nuclear power plants. He argued that an effective regulator does not silence its staff and noted that 47% of the NRC employees feel that it is unsafe to speak up to the NRC about problems at nuclear reactors. His recommendation— Congress must provide the attention and resources necessary to reform the NRC into a consistently effective regulatory body with a good safety culture.[34]
- When discussing the new Generation IV reactor designs, Lochbaum implied that while the industry is saying that they are safer than the old reactors, this is far from the case. These new designs utilize highly corrosive coolants under most extreme conditions of pressure and temperature, hence new super-resistant structural materials must be developed. Potential problems are further aggravated by the fact that some Generation IV reactors will be encased in sealed "batteries," with operating cycles which last from ten to thirty years, so that routine inspection and maintenance will be physically impossible. He also said that the exposure of new structural materials to extreme temperatures and powerful chemical reactions will be sure to create severe problems. His recommendation—experiments with new and untested materials must be conducted in laboratory

and prototype settings and not in commercial reactors that will operate near heavily populated areas.

- Some Generation IV reactors are to be cooled with liquid sodium, which burns and explodes when exposed to air or water, or with a lead-bismuth molten coolant, which is extremely corrosive and produces highly volatile isotopes when irradiated. Most Generation IV reactors are breeder reactors, which require reprocessing at the site—the so called "closed cycle." Reprocessing, while medically very dangerous because it releases large quantities of radioisotopes, requires the processing, transport, and storage of huge quantities of plutonium that can be made into nuclear weapons, which raises serious risks of proliferation and nuclear terrorism. Yet the United States is enthusiastically encouraging countries ranging from Brazil to South Korea to accept reprocessing facilities.

- Because of enormous safety problems most countries have abandoned reprocessing and breeders because they are exorbitantly expensive and dangerous. Lochbaum's recommendation—Congress must ensure that the next generation of reactor designs don't merely pursue but satisfy DOE's stated goals.

- The nuclear industry in its zeal to develop these exciting concepts has been arguing that Generation IV reactors will be so safe that emergency sirens and public protection measures will be unnecessary and should be eliminated because the emergency planning zone will shrink from 10 miles to a ¼ mile, thus eliminating the need for sirens.[35] At the same time, the industry has sought and gained federal liability protection under the amended Price Anderson Act to cover these ostensibly fool-proof new reactors. Lochbaum's recommendation—if the potential consequences of an accident at a Generation IV reactor are so catastrophic that plant owners require federal liability protection under the Price-Anderson Act, then emergency sirens and other emergency preparedness measures are a necessity for people living near those plants.[36]

The Future, or Merely a Replay of Nuclear's Past?

The U.S. government has taken the lead on Generation IV reactors. In the year 2000 the Department of Energy launched the "Generation IV International Forum" (GIF), which consists of 11 member nations including Argentina, Brazil, Canada, France, Japan, Republic of Korea, South Africa, Switzerland, the U.K., the United States of America, and Euratom—a consortium of European nations involved in nuclear energy. Not to be left behind, in 2001, the IAEA (International Atomic Energy Agency) initiated and funded a similar international forum, called the International Projects on Innovative Nuclear Reactors and Fuel Cycles (INPRO). Twenty-one countries are involved including Argentina, Armenia, Brazil, Bulgaria, Canada, Chile, China, Czech Republic, France, Germany, India, Indonesia, Republic of Korea, Pakistan, Russian Federation, South Africa, Spain, Switzerland, Netherlands, Turkey and the European Commission. (While there is some overlap in membership between these two organizations, the United States was reluctant to participate in IN-PRO because they view it as a Russian-inspired venture.)[37]

These international consortia provide large numbers of non-nuclear weapons nations with the equipment, expertise, and wherewithal to make nuclear weapons. This is an absolutely insane idea but one that seems to be proceeding apace with no public knowledge, oversight, or control.

Generation IV reactors will be so expensive that no single country possesses the necessary expertise nor the available funds to support the R&D. These new and innovative designs are so complex that they will be nowhere near completion until 2030 at the earliest with 2045 a more practical date—obviously they have not been conceived to make any difference to the global warming problem besetting the planet.[38] Global warming is happening right now. The situation is urgent and must be addressed at once. The money allocated to these new reactors could be used right now to mass produce renewable energy technologies that are currently available, and only these renewable technologies will have a positive effect to reduce global warming gases. As we know the nuclear fuel cycle substantially adds to global warming.

There is no uniformity of agreement, however, within the nuclear industry itself about which generation of reactors to pursue. While they all seem to support new reactor designs, a raging controversy is taking place, some favoring Generation III reactors, others favoring Generation IV.[39] Nuclear regulators in the United States are unenthusiastic about these new reactor concepts. According to one NRC commissioner, reactors should be based upon evolutionary, not revolutionary technology, because many of the safety problems inherent in these reactors are unresolved, many are unforseen and new problems will inevitably arise.[40]

In summary, the Generation III and Generation IV reactor designs are controversial and contentious, and seem not to be based upon sound economic, environmental, safety, or proliferation-resistant principles. To quote the skeptical Greenpeace report, "A revival of nuclear power is not to be expected—it will remain costly, is not competitive for hydrogen production and not suitable for developing countries. Is Generation IV a desperate attempt to get into hydrogen production despite all obstacles? Or is the goal simply to keep obsolete research installations running, which otherwise would be shut down due to safety concerns and lack of need? Is there a serious attempt to develop the HTGR technology, selling it as innovative while pursuing an evolutionary path? Or is it all only about an improvement of the image of nuclear energy, to be able to perform life extension of existing reactors while talking about Generation IV?"[41]

7

Nuclear Energy and Nuclear Weapons Proliferation

According to the British counterintelligence group MI5, over 360 private companies, university departments, and government organizations in eight countries, including Israel, Syria, Pakistan, Iran, India, Egypt, the Pakistani High Commission in London, and the United Arab Emirates, have been procuring nuclear technology and equipment for use in nuclear weapons construction. Companies in Cyprus and Malta may also be fronts for these weapons of mass destruction. MI5 states that the nuclear arms supermarket is larger than anyone realized and that front companies in the United Arab Emirates are the hub for the nuclear weapons trade.[1]

Because of the coming "renaissance" of the nuclear power industry, twenty-five countries and consortia will have access over a period of two decades to Generation IV reactors fuelled by plutonium. Some will also have "closed nuclear fuel cycles," with their very own reprocessing plants to produce plutonium, since the U.S. Department of Energy and the International Atomic Energy Agency (IAEA) in their wisdom, have involved these countries in their Generation IV International Forum—the International Projects on Innovative Nuclear Reactors and Fuel Cycles (INPRO). This diverse collection of states includes Argentina, Armenia, Brazil, Bulgaria, Canada, Chile, China, the Czech Republic, France, Germany, India, Indonesia, Japan, Pakistan, Republic of Korea, Russia, South Africa, Spain, Switzerland, the Netherlands, Turkey, the United Kingdom, and the United States, as well as the consortiums of Eurotom and the European Community. So at the same time that the IAEA and the U.S. government profess extreme concern about the proliferation of nuclear weapons, they are actively promoting and encouraging the dissemination of technology, expertise, and materials that make proliferation likely.

With sophisticated technology the minimum amount of plutonium required to make a bomb is 1 to 3 kilograms (2.2 to 6.6 pounds),[2] however the generally accepted amount is 5 kg of weapons grade plutonium and 8 kg for reactor grade plutonium. The design is available on the Internet; the essential materials can be bought at any hardware store. A homemade plutonium bomb would be difficult to make but a bomb using highly-enriched uranium would be less so. And the world is awash in plutonium. Russia and the United States each have 34 metric tons of plutonium accrued from the dismantling of nuclear weapons. If they continue dismantling more old nuclear weapons, they will accrue another 100 metric tons of free weapons-grade plutonium, while hundreds more tons of plutonium will remain in the nuclear arsenals of the world.[3]

Apart from the military plutonium, over 1,500 tons of plutonium has been produced by civilian nuclear power plants globally.[4] Although much of this material remains locked up in spent fuel rods mixed with highly toxic radioactive materials, countries such as Japan, France, India, Russia, and other European countries have been busily extracting their civilian plutonium from spent fuel rods by reprocessing their spent fuel. These countries have stockpiled over 200 metric tons at three large reprocessing plants: the Cogema facility in La Hague, France; British Nuclear Fuel's Limited Sellafield plant in Cumbria, England; and Mayak in Russia.[5] The Japanese are constructing a new reprocessing complex at Rokkasho on the northern island of Hokkaido;[6] and Germany, the Netherlands, Belgium, Switzerland, Italy, and the United States are among the other countries that already possess commercially separated plutonium.[7] If, as proposed by some, 2,000 new nuclear power plants are constructed over the next decades on the fallacious grounds of combating global warming, commercially produced plutonium could increase to 20,000 metric tons by 2050, dwarfing present inventories.

This is plutonium madness. Only one-millionth of a gram is a carcinogenic dose, and plutonium has a half-life of 24,400 years. In 1994, a report by the National Academy of Sciences described the Russian and U.S. military-derived stockpiles of plutonium as " a clear and present danger to national and international security," and a report by the British Royal Society in 1998 addressing the British stockpiles of plutonium concluded that "the chance that the stocks of (civil) plutonium might, at some stage, be accessed for illicit weapons production is of extreme concern."[8]

Unfortunately, although plutonium has been accumulating in vast stockpiles for many years, and at least five possible methods of plutonium disposal exist, to date not one laboratory or country has taken the most important essential steps to address the issue of loose plutonium. However most of these disposal methods leave much to be desired.

1. Plutonium could be mixed with uranium as MOX, or mixed oxide fuel, to be fissioned in light water civilian power reactors. This is not a good idea because it increases the amount of plutonium in civilian reactors, and, should there be a meltdown, there will be a large and very dangerous dispersal of plutonium to the four winds. Furthermore, the production of MOX fuel requires reprocessing of plutonium from weapons material, a very dangerous and dirty process.

2. Excess plutonium could be "transmutated" or fissioned in a fast reactor and converted to fission products that last 600 years, not 500,000 years, and that cannot be used for bomb fuel. But this does not solve the problem because these fission products are extremely radioactive, are medically dangerous, and concentrate avidly in the food chain compared to plutonium. However 600 years is still a

very long time in human terms—24 generations. Furthermore, most of the plutonium is converted to fission elements with 0.7 to 1% of plutonium remaining.[9]

3. Breeder reactors could use the plutonium to generate electricity while breeding more plutonium. This is an extremely dangerous technology as previously described.

4. The plutonium could be mixed with high-level nuclear waste, making it inaccessible to thieves because extremely intense gamma radiation would deter their intrusion.

5. MOX fuel, a mixture of plutonium and uranium, fashioned into ceramic pellets, could be placed in zirconium or stainless steel fuel rods, which could then be mixed with intensely radioactive spent fuel rods from reactors. This lethal combination could then be stored in massive cylinders placed in deep geological storage facilities. Thieves would have great difficulty accessing this material deep underground and would be heavily irradiated if they approached the rods.[10]

In light of terrorist attacks using conventional weapons, it is only a matter of time before someone steals enough plutonium to make an adequate nuclear weapon. Then we proceed into the age of nuclear terrorism.

Meanwhile, with the world awash in plutonium and highly enriched uranium, the Bush administration pursues its own nuclear armament development policy that makes it increasingly likely that a rogue nation will procure and possibly use nuclear weapons. The United States has adopted three contradictory stances at the same time:

- It is aggressively forging ahead to build more nuclear weapons, stating that it will use them preemptively even against non–nuclear nations.
- It is instrumental in denying the right to build nuclear weapons to all but a handful of countries.

• In the context of promoting nuclear energy, it has offered dozens of countries nuclear technology and access to nuclear power fuel. The fission process makes plutonium, which can then be separated by reprocessing and converted to fuel for nuclear weapons. While the Bush proposal includes taking the spent fuel back to the United States, it is not clear that that process can be undertaken with no cheating.

Thus, even as there is much hand-wringing at the United Nations about the possibility that Iran and North Korea may be developing nuclear weapons, eight nation-states—Russia, the United States, France, China, Britain, India, Israel, and Pakistan—possess their own nuclear arsenals, and others are free to develop weapons without the admonitions that the United States and the United Nations are imposing upon Iran and North Korea. This strange juxtaposition of opposing attitudes needs to be examined in the context of the sixty-five-year history of nuclear fission and related weapons development.

A BRIEF HISTORY OF NUCLEAR WEAPONS PRODUCTION

The term nuclear weapon encompasses several varieties of bombs, each of which employs different explosive mechanisms. An atomic bomb can be fueled by either plutonium or uranium. An atomic bomb works by either imploding its plutonium trigger with chemical explosives, which exerts tremendous symmetrical forces upon the plutonium, or by exerting huge pressures upon a mass of highly enriched uranium 235. The plutonium or uranium reaches critical mass, causing an explosion equivalent to the explosion of thousands of tons of TNT.

A hydrogen bomb is made of three components: a primary composed of an atomic bomb, which explodes first with a fission reaction; a secondary composed of deuterium and lithium, which then produces a fusion reaction similar to the reaction in the sun; and a tertiary mechanism produced when the uranium capsule of the bomb undergoes fission and explodes. A hydrogen bomb is relatively cheap for a country to build compared to deploying thousands of soldiers on the battleground, and the explosions can be of megaton range—equivalent to millions of tons of TNT. Most bombs today are hydrogen bombs.

America made three atomic bombs in 1945. The first was named Trinity after the Father, Son, and Holy Ghost and was exploded at Alamogordo in New Mexico in July 1945. The second, a uranium bomb called Little Boy, was exploded over Hiroshima on August 6, 1945, and the third, a plutonium bomb called Fat Man, was exploded over Nagasaki on August 9, 1945. Little Boy and Fat Man killed over 200,000 people, initiating the age of nuclear genocide.

The United States continued to make nuclear weapons after the end of the Second World War. Russia soon discovered the secret and joined the nuclear club in 1949; then Britain, France, and China got on board. In 1970, these five nations decided in theory that nuclear weapons should be abolished in the long-term and that, in the short term, only *they* should have the bomb; all others must be excluded from the nuclear club. To that end they drafted the Nuclear Non-Proliferation Treaty (NPT), which stated categorically that the nuclear nations would disarm and that non-nuclear weapons nations could not develop nuclear weapons. As compensation, the non-nuclear nations would be given access to "peaceful nuclear technology"—research reactors, nuclear power plants, and nuclear technology. The NPT, therefore, essentially gave non-nuclear countries the capacity to produce their own nuclear weapons, even as it forbade them to do so.

Under Article VI of the NPT, the nuclear armed nations also undertook not to enlarge their nuclear arsenals and to negotiate in good faith to secure their abolition. Since 1970 when the NPT was signed, the nuclear weapons nations have done the opposite, increasing their arsenals significantly.[11]

The overall state of world nuclear proliferation today is as follows:

- Eighteen countries now own uranium enrichment facilities which enable them to produce highly enriched uranium— the fuel for nuclear weapons. These countries include Pakistan, France, the United Kingdom, the United States, South Africa, Canada, Argentina, Brazil, Australia, China, India, Japan, Kazakhstan, and Russia.[12] It is not clear what uranium enrichment facilities Israel or North Korea now possess.
- Under the legal auspices of the NPT, seventy countries now have small research reactors, most of which are fuelled with highly enriched uranium, a fuel also suitable for nuclear weapons production.[13] These small research reactors also manufacture plutonium, making nuclear bomb materials available at each end of the research reactor's operation. Civilian nuclear power plants are mostly fuelled with low enriched uranium, unsuitable for nuclear weapons, but they manufacture plutonium—over 200 kilograms per year. And although some say that it is well nigh impossible to make a nuclear weapon from reactor-grade plutonium, in 1962 the United States tested such a nuclear weapon, and it worked very well.[14] Mohamed ElBaradei, the director of the International Atomic Energy Agency, is extremely worried about this situation and says that these widely distributed nuclear facilities are "latent bomb plants."[15]
- Nine countries now possess nuclear weapons, including the United States, Britain, France, Russia, India, Pakistan, Israel, China, and North Korea—an increase from the original five nuclear nations that signed the NPT.
- ElBaradei estimates that within a decade as many as forty more countries will have the ability to make nuclear weapons, and this may be an underestimate.[16]

- The United States has 10,500 nuclear weapons; Russia has 20,000; Israel has 110 to 190 or more; China has 400; France has 450; Britain has 185; India has 65; Pakistan has 30 to 50; North Korea has 2 to 9.[17]

Apart from the fact that so many countries now have access to nuclear technology, technologists from most of these countries were trained in nuclear technology by the nuclear-armed states, predominantly the United States. For instance, between the years 1955 and 1974, more than 1,100 Indian scientists received sophisticated nuclear training at U.S. nuclear facilities.[18]

In summary, seventy countries that now have the ability to develop their own nuclear arsenals are constantly being provoked as they observe the "nuclear club" refusing to disarm while the United States constructs even more nuclear weapons. Meanwhile, the United States and Russia still maintain thousands of nuclear weapons on hair-trigger alert, 2,500 on the Russian side and over 5,000 on the U.S. side. This means that hydrogen bombs are constantly mounted on their missiles, which are maintained in launch mode, and a command from the president of either country could launch a nuclear war within minutes.

Moreover, the early warning systems are now so tenuous that nuclear war could occur by accident and not by design. The Russian early warning systems are decrepit and failing, and none of their early warning satellites are operable so that they are blind most of the time. This is a very dangerous situation because the United States still maintains a first-use winnable nuclear war strategy against Russia. Consequently, the Russians are never sure when or if the United States will launch a nuclear strike and destroy them and their population in a nuclear holocaust.[19]

So serious in this situation that Boris Yeltsin in January 1995 came within ten seconds of pressing his nuclear button and destroying the United States by accident.[20] Because the missiles only take thirty minutes to go from launch to landing, and because it takes fifteen minutes for the nuclear command and control to determine whether or not the attack is real, both the Russian and the U.S. presidents are provided with only three minutes to make their launch decision.

With this situation as background, the Bush administration has adopted some very provocative and dangerous policies–all of them in direct violation of the Non-Proliferation Treaty—which inevitably have led and will continue to lead to the proliferation of nuclear weapons in other countries. For example, although the Cold War is over, a new semi-autonomous agency, the National Nuclear Security Administration, was established by Congress in the year 2000 within the Department of Energy to oversee the development and production of new nuclear weapons. Los Alamos National Labs has just produced its first trigger for an atomic bomb since the Cold War ended, a sphere of finely honed and lathed plutonium that nuclear scientists call a "pit." Los Alamos Labs have the capacity to produce 30–40 pits a year, and the plan of one DOE study is to make 500 new American hydrogen bombs annually, comparable to Cold War rates.

Because hydrogen bombs need tritium—radioactive hydrogen—for their fusion mechanism, the United States is also now proceeding to manufacture tritium at the Tennessee Valley Authority Watts Barr commercial nuclear power plant in Tennessee, which is then sent to the Savannah River site in South Carolina to be extracted by a new Tritium Extraction Facility. This is the first time since 1992 that tritium for nuclear weapons has been produced in the United States, and this is one of the few times that commercial and military enterprises have been combined in the United States.[21]

On the strategic front, the Bush administration has drafted a revised plan allowing military commanders to request presidential approval to use nuclear weapons to preempt an attack by a nation or terrorist group deemed to be planning to use weapons of mass destruction. These military commanders will also be permitted to use nuclear weapons to destroy known enemy stockpiles of chemical, biological, or nuclear weapons. The document says that preparations must be made to use nuclear weapons and to show determination to use them "if necessary to prevent or retaliate against WMD use." The United States has always had a "first-use" policy against nuclear-armed nations, but now this strategy is also being applied to non-nuclear nations for the first time. The "revised plan" reflects a preemptive nuclear strategy first enunciated by the White House in 2002.[22] Had this strategy been in place before the invasion of Iraq, a nuclear attack could have been justified to "take out" Iraq's imaginary weapons of mass destruction.

Other disturbing features of this document include authorization of the use of nuclear weapons against states without WMDs— to counter potentially overwhelming conventional adversaries, to secure a rapid end of a war on U.S. terms, or to "ensure success of U.S. and multinational operations." The draft document also gives the Pentagon permission to deploy nuclear weapons in parts of the world where their future use is considered most likely, and it urges troops to train constantly for nuclear warfare.[23] Although Congress voted in 2004 to discontinue research on the earth-penetrating nuclear bunker buster, this draft document calls for its continued development, and the U.S. Senate voted in July 2005 to revivify the bunker buster.[24]

With this combination of strategies and weapons developments under way in the world's most powerful nation, it is no wonder that many countries are pushing to develop their own nuclear arsenals. As Joseph Rotblat, an original member of the Manhattan Project who resigned from the project on moral grounds, said shortly before he died in August 2005, "If the United States, the mightiest country in the world, militarily and economically, feels that it needs nuclear weapons for its security, how do you deny this security to countries that really feel vulnerable."[25]

8

Nuclear Power and "Rogue Nations"

If the correct definition of a rogue nation is a state that possesses nuclear weapons and the ability to vaporize millions of people within seconds, eight or nine nations currently qualify: the United States, Russia, France, China, Britain, Israel, India, and Pakistan. North Korea may have two to nine nuclear weapons.

Because the United States and Russia possess the vast majority of nuclear weapons in the world—97% of the total arsenal of 30,000 bombs—and because these two countries continue to maintain thousands of these extraordinary weapons on "hair-trigger" alert, a nuclear exchange between them would kill billions of people and could induce nuclear winter and the end of most life, certainly in the northern hemisphere and much also in the southern hemisphere. Yet we persist in leaving them off the rogue roster.

As for those nations currently vying to add nuclear capability to their arsenals, nuclear power plants offer the perfect cover. It is only a short step from uranium enrichment for energy to the production of highly enriched uranium suitable for atomic bomb fuel, or even to reprocessed plutonium from spent fuel, suitable for bomb fuel. Most nuclear technology associated with nuclear power can be diverted for use in weapons production: North Korea has almost certainly built at least two nuclear weapons using plutonium obtained from its research nuclear reactors.

Many countries are angry about the paternalism and arrogance displayed over the years by the nuclear-haves. As the new president Mahmoud Ahmadinejad of Iran, which is now actively developing uranium enrichment facilities, said recently when referring to the United States, "Who do you think you are in the world to say you are suspicious of our nuclear activities? . . . What kind of right do you think you have to say Iran cannot have nuclear technology? It is you who must be held accountable."[1] Hugo Chavez of Venezuela displayed similar feelings when he said recently, "It cannot be that some countries that have developed nuclear energy prohibit those of the third world from developing it. We are not the ones developing atomic bombs, it's others who do that."[2]

In addition to Iran and North Korea, this chapter will look at three of the nuclear-haves who built their nuclear weapons arsenals using various components of the nuclear fuel cycle. Israel developed a very large nuclear arsenal from plutonium created in a reactor specifically designated for that purpose, India created a nuclear arsenal from heavy water nuclear power plants, and Pakistan developed nuclear weapons largely from uranium enrichment facilities.

IRAN

According to the rules of the Non-Proliferation Treaty, Iran is perfectly entitled to pursue a uranium enrichment program for peaceful purposes, in other words to manufacture uranium 235 enriched to 3% for use in a nuclear power plant.

In the past, Iran was actively encouraged by the United States to develop its own nuclear power program. And there was no shortage of companies willing to supply the means. The nuclear industry's drive to export nuclear reactors is not new: Tony Benn, a former member of the British Parliament, wrote in the *Guardian*, "Many years ago when the Shah—who had been put on the throne by the U.S.—was in power in Iran, enormous pressure was put on me, as secretary of state for energy, to agree to sell nuclear power stations to him. That pressure came from the Atomic Energy Authority, in conjunction with Westinghouse, who were anxious to promote their own design of reactor."[3] Had Iran built these reactors, Iran might possess nuclear weapons today, made from their nuclear power plant plutonium.

Now Iran is once again on the nuclear bandwagon. On September 17, 2005, the new Iranian president, Mahmoud Ahmadinejad, said in an address to the United Nations that Iran would not relinquish its right to pursue peaceful nuclear energy, and Russia is currently in the process of constructing a nuclear power plant at Bushehr in Iran.

Of course, one might wonder why Iran would want nuclear power when it is floating in oil, a question that could well have been asked when the Shah was in power. The genesis of Iran's uranium enrichment program began in 1985, during its war against Iraq, when it decided to pursue uranium enrichment even though there was no obvious need to provide fuel for a non-existent nuclear power program.

Iran later received advanced centrifuge designs in 1995 to enrich its uranium, but it is unclear when it actually started enriching the uranium—Iran claims not until 2002.[4] Iranian officials were not forthcoming in 2003, when inspectors from the International Atomic Energy Association discovered that Iran had been concealing efforts to enrich uranium; however, the International Institute for Strategic Studies said in September 2005 that Iran was at least five years from producing enough highly enriched uranium for a single nuclear weapon, and this would happen only if Iran ignored international condemnation and threw caution to the wind. (The report advised the international community "to apply international diplomacy in such a way that does not inspire Iran to abandon all restraint and seek a nuclear weapons capability without regard to international repercussions.")[5]

Inspectors to date have found no clear evidence to suggest that Iran is intent on creating nuclear weapons. Nevertheless, according to a September 2005 report by the International Atomic Energy Agency, despite an intense two-and-a-half-year investigation, key elements of Iran's nuclear program are shrouded in mystery. There is some new information on suspicious uranium contamination at some Iranian sites, and the agency stated that it is not in a position to say whether or not Iran is developing a clandestine nuclear program.[6]

Another report from the IAEA on November 18, 2005, said that Iran had provided it with a document that included partial instructions for developing the core of an atomic bomb. While the U.S. ambassador to the IAEA, Gregory Schulte, said that this disclosure raised concerns about weaponization, other diplomats and a U.S. nuclear expert, David Albright from the Institute for Science and International Security, were more cautious, saying that the instructions were far from a step-by-step guide to the production of a bomb core. The Iranians said that the document had been given to them unsolicited by people who were linked to the nuclear black market established by the Pakistani nuclear scientist Abdul Qadeer Khan.[7]

Despite the hazy nature of this evidence, somewhat similar to the flimsy evidence that the Bush administration presented about weapons of mass destruction before the Iraqi invasion, the United States has been adamant that Iran should be reported to the United Nations for failing to reveal all about its uranium enrichment program. Many nations belonging to the IAEA were extremely reluctant to take this move, but under extraordinary pressure by the United States and its staunch allies Britain, France, and Germany—and after an extremely rancorous debate, which was worse than any that anyone could remember—the board of the International Atomic Energy Agency finally voted 22 to 1 with twelve abstentions on September 24, 2005, to report Iran to the United Nations Non-Proliferation Treaty (NPT). (Russia, with a pending sale of five more commercial reactors to Iran,[8] and China, with extensive economic ties in Iran,[9] were persuaded to abstain rather than oppose the measure.)

The reluctance of the IAEA to refer Iran to the Security Council is explained by the fact that when Iraq was reported to the U.N. Security Council for possessing weapons of mass destruction, the British and Americans were provided with a legitimate excuse to invade Iraq.[10] And there are reports that the Bush administration is now preparing for a possible major attack against Iran, with extensive planning already underway at STATCOM, the United States Strategic Command, in Nebraska; the United States Central Command headquarters, in Florida; and the Joint Warfare Analysis Center, in Virginia. It is estimated that more than 400 strategic targets in Iran would need to be hit, including suspected underground nuclear weapons development sites.[11]

The plans for an attack on Iran call either for an air assault, and/or a combination of bombing and commando raids, or a full-scale proxy war conducted by Iranian opposition forces—the Mujaheddin-e Khalq (MEK), a group opposing the current regime.[12] Identification of possible targets began late last year when the CIA and U.S. Special Operations Forces (SOF) started flying unmanned "Predator" spy planes into Iran and sending small reconnaissance forces into Iranian territory to identify hidden Iranian weapons facilities. According to Seymour Hersh in *The New Yorker*, "The goal is to identify and isolate three dozen, and perhaps more, such targets that could be destroyed by precision (air) strikes and short term commando raids."[13]

Military analyst William Arkin said that it is probable that CENTCOM—U.S. Central Command—is probing Iran's air and coastal defense systems by sending electronic surveillance planes and submarines into Iran and its coastal areas. This exercise lights up their radars, identifying them for attack; the United States used this procedure when preparing for the Iraqi invasion in 2003.[14] This information is almost certainly being fed into strategic concepts and strike plans for an attack on Iran. On February 22, 2005, in Belgium, President Bush said: "This notion that the United States is getting ready to attack Iran is simply ridiculous. . . . Having said that, all options are on the table."[15]

Donald Rumsfeld, in the summer of 2004, approved a top-secret "Interim Global Strike Alert Order" to direct the U.S. military to maintain readiness to attack hostile countries, specifically Iran and North Korea, which are developing weapons of mass destruction, and this global strike includes the use of nuclear weapons. To this end, CONPLAN 8022-02, the Stratcom (Strategic Command) contingency plan completed in November 2003, calls for an offensive and preemptive strike capacity against Iran and North Korea with conventional or nuclear options.[16]

Having initially encouraged Iran to develop nuclear capabilities, the United States now has plans to bomb Iran with nuclear weapons for doing so. One may ask why the United States is so intent on invading Iran as it invaded Iraq. One answer might be Iranian oil. Another might be Israel. Pentagon officials have met with their Israeli counterparts to discuss the participation of Israel in plans to attack Iran, and Vice President Cheney confirmed in January 2005 that "the Israelis might well decide to act first," if Iran proceeded with the development of nuclear weapons.[17] In a predictably tragic twist, even if Iran turns out not to have been developing nuclear weapons, if the United States and/or Israel attack Iranian nuclear power facilities, huge amounts of radioactive material will be lofted into the air to contaminate the people of Iran and surrounding countries. This fallout will induce cancers, leukemia, and genetic disease in these populations for years to come, both a medical catastrophe and a war crime of immense proportions. Radioactive fallout will be hugely increased if the United States uses nuclear weapons on Iran's nuclear facilities.

In any event, the hypocrisy of taking a non-nuclear country to task for a lack of transparency regarding nuclear reactors is also clear. The United States and all other nuclear nations hide their nuclear activities with impunity, yet Iran may well be invaded for doing the same. As President Ahmadinejad himself pointed out in September 2005, "Every day they [the Americans] are threatening other nations with nuclear weapons, and they are never inspected." He said that Western countries were "relying on their power and wealth to try to impose a climate of intimidation and injustice over the world."[18]

NORTH KOREA

The situation regarding nuclear power and nuclear weapons in North Korea is at a happier pass as of November 2005 than it has been for some years, though the argument could be made that it was the existence of *research* nuclear reactors in North Korea, permitted under the Non-Proliferation Treaty (NPT) agreement, that led to a five-year standoff about nuclear *weapons* development in that country.

In March 1993, the IAEA stated that North Korea had produced more plutonium from reprocessing its spent fuel than it had disclosed. North Korea, annoyed by this seeming intrusion on their honesty and national sovereignty, announced that it would withdraw from the NPT. The IAEA inspectors are allowed to enter a country *only* to ascertain whether it is diverting plutonium or highly enriched uranium into weapons. Washington believed at that time that North Korea had extracted enough plutonium for two crude nuclear weapons.[19]

This impasse was solved when bilateral negotiations were arranged between the Clinton administration and North Korea. On

October 2, 1994, they drew up an Agreed Framework, announcing that North Korea would freeze its nuclear program—close its five-megawatt Yongbyon reactor and stop work on building two plutonium–producing graphite reactors. In exchange, North Korea would receive two light water nuclear power plants, to be constructed by an international consortium. In addition, prior to construction of the reactors, they would also receive 500,000 tons of heavy heating oil annually to help them through their very cold winters.[20]

From the beginning, however, the second Bush administration adopted a hostile and provocative stance toward peace and reconciliation initiatives on the Korean Peninsula. Important internal diplomatic initiatives had taken place on June 15, 2000, when South Korean President Kim Dae-jung and North Korean President Kim Jong-Il signed a peace treaty stating the need for reconciliation and peace. However, when the South Korean president, a Nobel Peace Prize laureate, paid a visit to George W. Bush soon after Bush's inauguration, he was essentially snubbed by Bush and his staff when nobody would meet with him when he visited the White House at the very time that he was continuing his delicate negotiations with North Korea to open up venues of trade, communication, trust, and exchange of citizens between North and South Korea.[21] This translated into an insult to North Korea as well. President Bush compounded the problem in his 2002 State of the Union address when he belligerently accused North Korea, Iran, and Iraq of being part of an "axis of evil."[22]

Ultimately, the Agreed Framework collapsed in October 2002, when the Bush administration accused North Korea of producing enriched uranium for nuclear weapons. Yet the Agreed Framework specifically stated that North Korea would dismantle its plutonium-producing graphite reactors and suspend other related activities *only after* the light water reactors were completed. Although a groundbreaking ceremony for the reactors had taken place, the light water nuclear power plants had not yet been built; the North Koreans were operating within the Agreed Framework.[23]

In 2003, in retaliation for these provocative acts, North Korea reopened their five-megawatt research Yongbyon reactor, expelled the IAEA inspectors, and announced that they had reprocessed spent fuel to obtain plutonium.[24] The Bush administration then set out to terminate the Agreed Framework, to remove themselves from bilateral negotiations established under Clinton, and to stop North Korea from reprocessing spent fuel. But the international community, worried that North Korea would retaliate for these provocative moves on the part of the Bush administration by developing an arsenal of nuclear weapons, put heavy pressure on the United States to include other nations in the negotiations. After initial resistance from the Bush administration, a six-party negotiation with North Korea was organized, involving China, Japan, Russia, and South Korea, as well as North Korea and the United States. The first meeting involving the six nations was held in August 2003. This meeting, however, was seen as a contentious move by the North Koreans. Their goal had been to negotiate a non-aggression pact directly with the United States.

Negotiations between these six nations dragged on for several years, with the United States and North Korea still fundamentally at loggerheads. The U.S. delegate, the former anti-proliferation official John Bolton, employed confrontational tactics and abusive name-calling. Ultimately, however, shortly after being appointed secretary of state, Condoleezza Rice stepped in and appointed Assistant Secretary of State Christopher Hill to take over the negotiations. President Bush then stopped calling President Kim Jong Il a tyrant, and abandoned his "axis of evil" rhetoric.[25]

These moves did not obliterate the animosity, as North Korea continued posturing in a belligerent fashion, intermittently retreating from the talks. Things looked especially grim in February 2005, when North Korea announced to the world that it had

nuclear weapons for self-defense and that it would not return to
the talks unless the United States ceased its hostility.[26]

But amazingly, persistent and professional negotiation and
diplomacy led to a mutually acceptable position on September 19,
2005, when the six parties reached agreement:[27] North Korea
would give up all nuclear weapons and existing nuclear weapon
programs in exchange for a civilian nuclear power plant, economic
aid, security commitments, a possible normalization of relations
with the United States, and a large infusion of electricity from
South Korea. These benefits were contingent upon North Korea's
rejoining the Nuclear Non-Proliferation Treaty and readmitting
the IAEA inspectors.[28]

Significantly, the United States and North Korea agreed to respect each other's sovereignty and right to peaceful coexistence and
to work toward a normalization of relationships. This latter part of
the agreement is very important, because these two countries do
not have full diplomatic relations and did not sign a peace treaty
following the end of the Korean War.

ISRAEL, INDIA, AND PAKISTAN

Israel, India, and Pakistan are the only countries in the world with
nuclear capabilities not to have signed the Nuclear Non-Proliferation
Treaty (NPT). They have developed their own clandestine nuclear
arsenals, and they have never been subject to IAEA verification inspections. As such, they are truly rogue nations, outlaws who choose
not to abide by international law.

Israel

Israel developed its nuclear arsenal using plutonium manufactured in
a heavy water nuclear reactor called Dimona built in the Negev
desert. France provided the bulk of the assistance to build this reactor,

and it went on-line in 1964. This reactor was specifically a bomb factory, not a power plant.

Israel has a nuclear arsenal variously estimated between 100 and 400 nuclear weapons, according to many experts, although it has repeatedly refused to confirm or deny that it possesses them.[29]

On September 28, 2005, Egypt put forward a proposal at the Vienna meeting of the IAEA that the Middle East become a nuclear-free zone. The Egyptian resolution stated that "we have one state in the area that constitutes an exception in Israel which remains outside the NPT regime and any legal framework in the area of nuclear disarmament.... To build confidence . . . you must have one element . . . renounce possession of nuclear weapons, create an area free of weapons of mass destruction [and agree to] full verification on the part of the IAEA."[30] This was a futile attempt by Egypt to get Israel to acknowledge its nuclear arsenal and to submit to IAEA inspections.

Ultimately, this effort demonstrated the futility of trying to get Israel to conform to the international laws imposed with such stringency upon other states such as Iran and North Korea. Predictably, Israel refused to entertain this proposal. Understandably, the Arab states resent the IAEA's intrusions on Iran, as the United States accuses it of a covert but un-proven nuclear weapons program, whereas Israel, also a covert nuclear state but a close U.S. ally, receives no such scrutiny. The chief of the Israeli atomic energy commission cynically said that the Arab initiative was "politically and cynically motivated"[31] and was an attempt to name Israel as a nuclear nation.

In any event, it is unwise and dangerous for Israel to possess a nuclear arsenal. Such weapons are highly provocative for Israel's Arab neighbors. And they are also dangerous to Israel, because the presence of Israeli bombs actively encourages Arab states to build their own. One or two nuclear bombs landing on the tiny Israeli nation would obliterate it. Or, conversely, if a large conventional weapon landed on Dimona, the ensuing meltdown would kill millions of people.

Nuclear nations, including Israel, need to take responsibility for their arsenals and start disarming with serious intent, abiding by Article VI of the Nuclear Non-Proliferation Treaty.

India

India created its first nuclear weapons using plutonium made in a research nuclear reactor, later moving to plutonium manufactured in nuclear power plants. India, too, is a rogue nation. By choosing not to belong to the NPT, it is closed to international inspection that the world insists on for Iran and North Korea. Furthermore, an unholy nuclear alliance called the Indo-U.S. Joint Statement has developed between India and the United States. An agreement signed on July 18, 2005, between President George Bush and Prime Minister Manmohan Singh gives India help for its civilian nuclear power program while allowing it to maintain its nuclear arsenal outside the NPT.

In a classic case of beating ploughshares into swords, India developed nuclear weapons out of nuclear power materials provided by Britain, Canada, and the United States. Britain provided design details and enriched uranium for India's first research reactor at Apsara. Canada supplied India with a CIRUS heavy water reactor for making nuclear energy, and the United States supplied the heavy water. It was this reactor that gave India the plutonium it used in its first 1974 nuclear weapons test.[32]

The United States then gave India its first commercial nuclear power plant at Tarapur, and Canada provided the second at Rawatbhata. Despite enormous economic outlays, nuclear power provides only 3% of India's total electricity, less than India's rickety old windmills. India's reactors are the most contaminated in the world, exposing hundreds of workers to excessive doses of radiation.[33]

A scientific study of the health consequences on the local population adjacent to the Rajasthan nuclear power plant found statistically significant increases in the incidences of congenital deformities, spontaneous abortions, stillbirths, and solid tumors.[34]

India, using these power plants for creating both nuclear energy and fuel for nuclear weapons, ultimately constructed sixty-five nuclear weapons while consistently calling for a nuclear-free world. Over 1,100 of India's nuclear scientists were trained in the United States.[35]

The Indo-U.S. Joint Statement sanctions the nuclear arsenal of India even though the country has not signed the NPT, but says that India must identify and separate civilian and military nuclear technologies and open up its civilian facilities to IAEA inspections for the first time. By contrast, France, Britain, Russia, and China have never separated their civilian and nuclear weapons programs. And the United States has placed only a tiny handful of its nuclear reactors under IAEA safeguard inspection. The safeguards regime has hypocritically assumed that these five nuclear-weapons powers have the right to divert civilian materials to military uses because they are the "legitimate" possessors of nuclear weapons.[36]

Under the terms of the agreement, India must not test nuclear weapons, although this testing moratorium will be tenuous because the United States is setting a bad example. The US Senate has not ratified the Comprehensive Test Ban Treaty, and the United States is already performing subcritical weapons testing in Nevada and is predicted to start actual nuclear weapons tests soon.[37]

Under the Indo-U.S. Joint Statement, the United States states that it will work with its friends and allies to provide expeditious fuel supplies to the Tarapur reactor, an agreement that may however be illegal under U.S. congressional rules. A 1978 law bars American nuclear energy aid to nuclear weapons states, and the coalition of nations called the Nuclear Suppliers Group adheres to similar restrictions.[38]

However, this deal will encourage uranium suppliers from other countries to sell to India, thus freeing up India's domestically mined uranium to be diverted to its weapons program, which will be unpoliced. Such a laissez-faire arrangement will allow India to stockpile enough plutonium by 2010 to make another 130 nuclear weapons.

Furthermore, India's commercial spent fuel has not been placed under safeguards. India now has 8,000 kilograms of reactor-grade plutonium in their spent fuel—enough to make another 1,000 nuclear weapons. If it is not now carefully policed by the IAEA inspectors, India could become the third-largest nuclear weapons state in the world (behind the United States and Russia).[39]

So what are the real reasons behind the United States' dangerous agreement with India? The surreptitious agreement was signed behind closed doors between Dr. Singh and President Bush without the oversight of the Indian cabinet, its committee on security, the U.S. National Security Council, the U.S. National Security Advisory Board, or even the U.S. Department of Atomic Energy. The only advance briefing in India about the secret U.S.-Indian deal was delivered to select members of the media by the U.S. embassy in India.[40]

The United States wants to accelerate India's rise as a global power to act as a regional counterweight to China. U.S. business plans intersect with strategic plans for India. When President Bush signed the joint statement, he said that, "as a responsible state with advanced nuclear technology, India should acquire the same benefits and advantages as other such states." These "benefits and advantages" signify that India would buy $15 billion worth of conventional military equipment from the United States including anti-submarine patrol aircraft to spot Chinese submarines in the Indian Ocean and Aegis radars to assist Indian destroyers operating in the strategic Strait of Malacca to monitor the Chinese military.[41]

Under the agreement, India may also purchase the Arrow missile system, developed by Israel with American technology, and the new AP-1000 nuclear power reactors made by Westinghouse. In this context, Dr. Singh, the Indian president, has also called for private investment in Indian nuclear power generation—a move that could open the door for U.S. companies to hawk their Generation III and IV nuclear reactors to India, with serious repercussions to India's strategic interests, national security, sovereignty and independence.[42]

In summary, the main ingredients of the U.S.-Indian agreement are:

- The United States legitimized India's clandestine nuclear weapons program, setting a new and dangerous precedent that will justify clandestine nuclear programs in countries that have not signed the NPT or who are involved in the illegal production of nuclear weapons.
- The United States is imposing upon India the job of "containing" the Chinese, a move that could reignite old tensions that continue to simmer between India and China, and possibly Pakistan.
- The United States is encouraging India to buy massive quantities of military equipment, although millions of its people survive at a barely subsistence level.
- The United States is encouraging India to purchase new nuclear reactors and involving it in the development of the dangerous Generation IV reactors.

This agreement will seriously undermine U.S. and U.N. efforts to confront possible illegal weapons programs in North Korea and Iran. Will the United States next be offering nuclear power technology and advanced conventional weapons systems to countries such as Brazil, South Africa, South Korea, Taiwan, and others who may also be producing clandestine nuclear weapons?[43] This agreement effectively destroys the legitimacy of all international nuclear safeguard agreements, so carefully negotiated by the international community and the United Nations.

Pakistan

Pakistan has utilized various elements of its nuclear fuel cycle to create nuclear weapons—in large part from its uranium enrichment facilities and possibly from plutonium obtained from its research reactor.

The illegally nuclear-armed nations Pakistan and India came exceedingly close to nuclear war in 1999 during fierce fighting over Northern Kashmir. Their nuclear-armed missiles, which take several minutes to reach their target, remain on hair-trigger alert and can be launched at any time with the press of a button. Neither side retains an adequate early warning system.[44]

Their mutual animosity is ancient and simmers along, despite some moves at reconciliation. Pakistan began its nuclear weapons program in 1972 under the leadership of Prime Minister Zulfiqar Ali Bhutto, but picked up the pace after India tested its first nuclear weapon in 1974. In 1975, a German-trained metallurgist named Dr. Abdul Qadeer Khan returned to Pakistan from the Netherlands with a knowledge of gas centrifuge technologies used for enriching uranium, along with some stolen uranium enrichment technologies from Europe.[45]

Khan was given the mandate to build, equip, and operate the Kahuta nuclear facility in Pakistan, specifically designed to enrich uranium. An extensive clandestine network was established to obtain the necessary materials and technology to develop Pakistan's uranium enrichment facilities.[46]

By 1985, Pakistan was producing weapons-grade uranium, and by 1986 it had produced enough enriched uranium for a nuclear weapon. In 1987, it exploded its first atomic bomb, and by 1998 it had conducted six nuclear weapons tests. These tests took place two weeks after India had conducted five nuclear tests, and after Pakistan warned that it would respond to India's tests.[47]

The creation of the Pakistani bomb was aided and abetted by China, which played a major role in the development of Pakistan's nuclear infrastructure at a time when Western countries were increasingly stringent on nuclear exports and assistance. In the 1990s, China supplied Pakistan with the heavy water for the Khusab research reactor, which produced weapons-grade plutonium, and the design of one of Pakistan's nuclear weapons. China also supplied components for Pakistan's high-speed uranium centrifuges, technical assistance and materials for their Chasma nuclear power reactor, and assistance to construct the Pakistani reprocessing facility in the 1990s. The former Soviet Union and Western Europe also contributed to Pakistan's nuclear weapons program by supplying dual-use nuclear equipment.[48] As a result, Pakistan is now the proud owner of thirty to fifty nuclear weapons, despite the fact that it has not signed the Comprehensive Test Ban Treaty, and it maintains a first-use policy.[49]

After years of weapons build-up on both sides, Indian Prime Minister Atal Bihari Vajpayee and Pakistani Prime Minister Nawaz Sharif signed the Lahore Agreements in February 1999, which mandated confidence-building measures and a continuation of a bilateral moratorium on nuclear testing.[50] But these diplomatic advances were undermined soon after when Pakistan, in a provocative act, invaded Kargil in the much-disputed area of Kashmir. Under U.S. pressure, Prime Minister Sharif withdrew his troops; he was subsequently deposed in October 1999 by a military coup led by General Pervez Musharraf.[51]

But the subliminal rivalry between the two countries is never far from the surface on the Indian subcontinent. General Musharraf recently said that although Pakistan does not want a conflict with India, if it came to war between these two nuclear rivals, he would respond with "full might."[52]

In the meantime, Dr. Khan, the importer of nuclear technologies, became a national hero, having been responsible for Pakistan's pride in its nuclear weapons program. And Khan did not confine his expertise to Pakistan; his hubris was such that he established the largest nuclear smuggling ring in history, supplying non-nuclear states with the technology, equipment, and ability to make nuclear weapons. In January 2004, he confessed publicly to having provided nuclear weapons technology to Iran, North Korea, and Libya. For these grave international nuclear crimes, Dr. Khan is now under house arrest in Islamabad, but he remains a national hero, and President General Musharraf granted him a pardon.[53] There are statues commemorating Khan in many locations in Pakistan.

The history between the United States and Pakistan is interesting and variable. Several times during its nuclear buildup, the U.S. imposed trade and military sanctions on Pakistan because of its clandestine weapons program, but the sanctions were suspended when the United States needed Pakistan as a strategically important ally.[54] On the nuclear front, the United States has consistently failed to come down hard upon Musharraf because he is desperately needed as an ally.

When the Soviet Union invaded Afghanistan on December 24, 1979, the United States sent weapons, military training expertise, and intelligence services through Pakistan to the majahadeen, the Taliban, and Osama bin Laden to fight the Russians. However, after the 9/11 attacks in 2001, America quickly changed sides and exerted pressure upon Pakistan to fight its former allies in Afghanistan.

This is a very volatile area of the world, and although the United States maintains a close alliance and friendship with President Musharraf, many members of his military belong to al Qaeda and the Taliban and are dissatisfied with the current alliance. Should there be a successful coup against Musharraf, these military people would then gain possession of Pakistan's nuclear arsenal, which they could rapidly share with the Taliban and al Qaeda.[55]

The use of Pakistani nuclear weapons could trigger a chain reaction. India, an ancient foe, would almost certainly respond in kind. China, India's hated foe, could then react, triggering a nuclear holocaust on the subcontinent, killing millions of people.[56]

INTERNATIONAL APPROACH TO NON-PROLIFERATION

The United States is directly responsible for the breakdown of the Nuclear Non-Proliferation Review conference at the United Nations in May 2005, which collapsed when U.S. Under Secretary for Disarmament John Bolton refused to participate in meaningful discussions, thereby sabotaging the meeting, to the despair and disgust of the rest of the world.[57] It was the collapse of the NPT review that induced Norway, Britain, Australia, Indonesia, Chile, and Romania to propose new rules on nuclear weapons proliferation and disarmament. International negotiators worked for months on a reform package to present at the United Nations summit in September 2005.[58]

However, by the time of the summit, John Bolton had been promoted to the position of U.S. ambassador to the United Nations, and he intervened, vehemently objecting to the United Nations' focus on disarmament of the major powers, rather than on the spread of nuclear weapons among rogue states and terrorists. Bolton intervened only days before the summit began, and because the United States wields enormous power at the United Nations, the proposed new rules on disarmament and nuclear weapons proliferation were completely disregarded.[59]

U.N. Secretary-General Kofi Annan concluded the summit by ruing its outcome. He made a final speech in which he said, "It's a real disgrace," lamenting the omission, blaming "posturing" for a failure to find a common approach to the spread of weapons of mass destruction. He said that nuclear non-proliferation and disarmament is "our biggest challenge" and our "biggest failing," as he recalled the collapse of the NPT review conference earlier in 2005.[60]

9

Renewable Energy: The Answer

The good news is that there is no need to build new nuclear power plants to provide for the projected energy needs of the future. Indeed, it would be possible, using other forms of electricity generation, to close down most of the existing nuclear reactors within a decade. There is enough wind between the Rocky Mountains and the Mississippi River alone to supply three times the amount of electricity that America needs.

Many kinds of alternative solutions are currently on the drawing board because of the extreme urgency of countering global warming. For instance, the conversion of coal to a synthetic fuel, which can be used for transportation and which would contribute much less to global warming than petroleum, is actively being championed by Governor Brian Schweitzer of Montana.[1] This chapter, however, concentrates exclusively on *renewable* sources of power for the generation of electricity. The most commonly cited figures show that currently in the United States, just over 2% of the electricity is provided from renewables, whereas nuclear power provides 20%.[2] These figures, however, exclude hydropower electricity. If this is taken into account, 2004 figures show 9.06% of U.S. electricity came from renewables, and 18.60% came from renewables worldwide.[3]

But American politicians lack the political will, at least at the federal level, to resist the coal, oil, and nuclear industries' demands and to shift their focus from these tired and dangerous technologies to explore the alternatives. Vice President Cheney devised the 2005 energy bill behind closed doors, consulting exclusively with top executives of the coal, oil, and nuclear industries (including Ken Lay from Enron who is currently under indictment), all of whom had contributed significant funds not only to the Bush campaign, but also to the campaigns of most of the important Republican players in the House and Senate. Thus, American politicians are bought and sold, and global warming continues unabated.[4]

However, the world at large has already begun to shift over to alternative energy sources, as is documented in several recent studies. A 2005 Rocky Mountains Institute report by Amory Lovins titled "Nuclear Power: Economics and Climate-Protection Potential," uses industry and government data to show that globally, nuclear power is being outstripped by other, better sources of electricity production. Globally, more electricity is now produced by decentralized, low-carbon or no-carbon competitors than from nuclear power plants—about one-third from renewables (wind, biomass, solar) and two-thirds through a very efficient form of energy production in which electricity is made from waste heat emanating from industry in a process called fossil-fuel combined-heat-and-power CHP, or co-generation.[5]

Even without the subsidies enjoyed by the nuclear industry, worldwide, decentralized electricity generators provided almost three times as much output and six times as much capacity as nuclear power by 2004.[6] (Output is the actual amount of electricy generated, whereas capacity is the *potential* output of an electricity generator. These two numbers are different whenever generators are not operated at top capacity.) And decentralized capacity is projected to increase 177 times by 2010, at the same time that orders for new nuclear reactors decline and aging reactors shut down. Nuclear power plants take years to build, they are energy intensive, and they are extremely expensive. Lovins contends that even the relatively inefficient use of decentralized electricity generation in today's market supersedes nuclear electricity in cost, speed, and size by a large and rising margin.[7]

Lovins ultimately concludes that none of the centralized thermal power generators (coal, gas, oil, or nuclear) can compete economically with wind power and certain other renewables, let alone the two cheaper alternatives (cogeneration and energy efficiency).[8] He finds it interesting that most of the studies that examined the energy future, such as the oft-quoted 2003 MIT study, fail to examine the feasible economic alternatives to nuclear and large centralized generation.[9] And, as stated above, the U.S. administration and Congress apparently have no intention of seeking the obvious economic and ecological alternatives to coal, oil, and nuclear power.

According to Lovins, the oft-made claim by nuclear energy proponents that "we need all energy options" has no analytical basis and is simply not true. Quite the contrary, society cannot afford all options. Because the disastrous economics of nuclear power mitigate against private investment, all new orders for nuclear reactors are to be heavily subsidized by taxpayers—$13 billion is allocated to the nuclear industry in the 2005 U.S. energy bill, for example.[10] Although a bonanza for nuclear power plant owners, this money is directly diverted from investment in cheaper, cleaner, greener options—cogeneration, renewables, and efficiency—that would ultimately serve consumers and the environment infinitely better.[11]

THE ECONOMICS OF ALTERNATIVES

The 2003 MIT study on the future of nuclear power demonstrates that each ten cents spent to buy a single nuclear kilowatt hour (kWh) of electricity could instead be used to generate 1.2 to 1.7 kWh of gas-fired electricity, 2.2 to 6.5 kWh of cogeneration from large industries, an infinite number of kilowatt hours of waste heat cogeneration, or 10 kWh of saved energy through efficiency measures.[12]

The New Scientist, a well-known scientific journal published in the United Kingdom and the United States, recently editorialized that although renewable electricity technologies are heavily criticized by the nuclear, coal, and oil industries and by many politicians who listen to the industry propaganda, the combination of wind power, tidal power, micro-hydro, and biomass make renewable power ever more practical. Windpower and biomass are now almost as cheap as coal, and wave power and solar photovoltaics are rapidly becoming competitive.[13]

A report from the New Economics Foundation reinforces the conclusions of the *New Scientist*. Renewable energy is quick to build, abundant, and cheap to harvest, and it is safe, flexible, secure, and climate friendly. Surplus electricity can be fed back into the grid. Furthermore, renewable electricity generation produces electricity at the point of use, making large-scale grid connections unnecessary. Thus, from an economics standpoint, renewable sources of energy make a great deal of sense.

THE ENVIRONMENTAL IMPACT OF ALTERNATIVES

To have any impact upon carbon emissions that contribute to global warming, fast and effective climate solutions must be implemented immediately. Empirical data confirms the terrifying phenomena of global warming:

- At least twenty severe catastrophic weather events occurred every year since 1990. During the preceding twenty years, only three such years were recorded.
- The four hurricanes that devastated the United States in 2004 produced a record loss of $56 billion over a period of several weeks.

- In 2004, ten typhoons hit Japan, four more than any previous record, the costliest year ever caused by Japanese typhoons.
- Europe experienced its hottest summer on record in 2003, which killed 22,000 people from heat-related illnesses and caused catastrophic wildfires incurring $15 billion in losses.
- The number of severe winter storms doubled in Britain over the last fifty years.[14]

The U.S. administration, however, has long attempted to ignore or to deny that global warming even exists. President Bush refused to sign the Kyoto Protocol on climate change in 2001,[15] and the United States secretly undermined Prime Minister Tony Blair's attempts to tackle climate change at the G8 Gleneagles meeting in 2005. Leaked U.S. government documents associated with that meeting reveal that the science of climate change was subverted, and that the United States had withdrawn from a crucial United Nations commitment to stabilize greenhouse gas emissions. Washington officials had:

- removed the reference to climate change as a "serious threat to human health and to ecosystems";
- deleted suggestions that global warming is happening;
- deleted suggestions that human activities are responsible;
- deleted the statement that "Africa, Asia–Pacific and the Arctic are particularly vulnerable to climate variability and are starting to experience the impacts";
- reneged on economic pledges to provide a network of regional climate centers throughout Africa designed specifically to monitor the ongoing impact of global warming.[16]

Unlike nuclear energy, all the renewable forms of energy mentioned above are extremely effective carbon displacers per dollar. For example, each $100 spent to develop nuclear power instead of end-use efficiency translates into the atmospheric release of one ton of CO_2. Nuclear power thus contributes to global warming by diverting valuable assets away from all environmentally sounder alternatives such as wind power, solar power, geothermal energy, biomass, and cogeneration, each of which produces very little if any carbon dioxide.[17]

THE POLITICS OF ENERGY IN AMERICA

Much to everyone's surprise, some weeks after Hurricane Katrina, President Bush began to encourage Americans to save energy, to drive less, to observe the speed limit, and to buy smaller cars, while ordering the federal government to conserve energy. On October 3, 2005, the Department of Energy announced a campaign to convince Americans to use less energy. The DOE even produced a new mascot called Energy Hog, like Smokey the Bear, to encourage people to conserve. They also published booklets with conservation tips, established a Web page, distributed public service announcements to radio and newspapers encouraging energy conservation, and sent specialists to factories to advise on energy conservation. Astonishingly, the Alliance to Save Energy, which is partnering with the DOE, officially announced that America is wasting half its fuel.[18]

Maybe this signals a significant change in the Bush administration in light of its previous record. Perhaps they too are getting worried about global warming and not just the shortage of oil. Paul Krugman, however, writing in the *New York Times*, is skeptical. He says the Bush administration is just responding to public anger over higher heating oil and natural gas prices secondary to Hurricanes Katrina and Rita. Krugman argues that there is little to suggest a substantial shift from Bush's "drill-and-burn" energy policy, pointing out that the administration has made no significant changes, such as raising mileage requirements for cars and trucks, or raising gasoline taxes. (Australia and many other countries pay up to three times as much for their gas as the United States, which has always maintained artificially low, subsidized prices.)[19]

But more substantial developments are taking place at the state level in the United States. Many states are banding together in regional agreements to regulate carbon dioxide emissions. The first regional agreement announced in the Northeast in September 2005 plans to freeze CO_2 emissions from big power stations by 2009 and then reduce them by 15% by 2020. This large industrial area stretching from New Jersey to Maine generates approximately the same volume of CO_2 emissions as Germany. Participating states include New Jersey, New York, Massachusetts, Connecticut, New Hampshire, Maine, Vermont, Rhode Island, and Delaware.

The second regional agreement is evolving on the West Coast, as California, Oregon, New Mexico, Washington, and Arizona explore similar pacts. Increasingly, other states are attracted to these measures.[20]

These new initiatives are especially significant because the United States continues to retain its status as the world's most profligate consumer of energy. For instance, in 2001, 49% of the electricity-related carbon dioxide emissions came from the industrialized world—North America, Western Europe, and industrialized Asia, of which U.S. emissions were 24%, which accounts for a mere 4.5% of the world's population, whereas the other 51% of CO_2 emissions emanated from the developing nations, the former Soviet Union, and Eastern Europe.[21]

TWO EXCELLENT ALTERNATIVES

Wind Power

Wind power, already used extensively in Europe, is rapidly becoming the energy of the future. It is cheap, fast to produce, and attractive to farmers and U.S. rural communities. In 2004, wind power

globally outpaced nuclear power sixfold in annual capacity additions and threefold in annual output additions. Wind power is very
attractive because it is benign, its development has short lead times,
its mass production is economically very efficient, its technological
development is rapid, and it is easy to site windmills on available
land. Furthermore, the speedy deployment and lack of regulatory
fuss will always support the growth of wind power compared to
the long lead time and delay-prone, complex, and contentious
technology of nuclear power, which could experience a meltdown
or terrorist attack at any time.[22]

A recent study, which collated more than 8,000 wind records
from every continent, found a potential global wind power resource of 72 terawatts—forty times the amount of electricity used
by all countries in 2000. If just 20% of this wind energy were to be
tapped, all energy needs of the world could be satisfied (one ter-
awatt of electricity would power 10 billion 100-watt light bulbs).[23]
(This analysis of global available wind power performed by
Christina Archer and Mark Jacobson of Stanford University is
probably somewhat conservative in scope because many continents
lack the specific data for wind over large unmapped areas.)

The most powerful wind forces in the world occur in the
North Sea in Europe, the Great Lakes of North America, the northeast and northwest coasts of North America, and the southern tip
of South America.[24] Archer and Jacobson found that, although
wind generation has increased at a remarkable rate of 34% annually
over the last five years, becoming the fastest growing source of
electricity production, wind currently provides a mere half percent
of the world's energy.[25]

Stimulated by the world's oil crisis in the 1970s, Denmark decided to develop wind energy. In 1988, two years after the Chernobyl accident, the Danes passed a law forbidding the construction
of nuclear power plants. This country is now the world's leader in
a large, lucrative wind energy technology and is pursuing the

fourth generation of wind turbines. Most Danes are delighted with their decision. As one of them said, "I wanted my children to have five fingers, we made a choice: No nuclear energy. We're going to do something else." (Sadly, the legacy of Chernobyl will linger for hundreds of years in Denmark, as tragically some of their food is still radioactive.)[26]

Wind power has enormous potential in the United States. The land between the Rockies and the Mississippi is referred to as the Saudi Arabia of wind because of the relentless gales that consistently batter this huge American prairie.[27] Texas, Kansas, and North Dakota together could provide 100% of America's electricity. The offshore potential for wind energy is incalculable, and the wind potential of the Great Lakes and the northwest and northeast areas of the States has hardly been tapped.[28] Wind power from readily available rural land in just several Dakota counties could produce twice the amount of electricity that the United States currently consumes.

In Minnesota, since the mid-1990s, hundreds of wind turbines have been generating electricity through this windswept region. Constructed by large corporations who pay farmers $2,000 to $5,000 per machine to rent their land, wind power machines have produced enormous benefits to cash-strapped farmers. Some farmers have even developed their own commercial-scale, giant wind turbines on wind-farms called "combines in the sky," making even more money from this new, green energy crop.[29]

At the governmental level, several problems need to be addressed for the wind power industry to reach an appropriate scale. While wind power generated in the U.S. Northeast has ready access to electricity grids, which are plentiful in that part of the country, access to the grid is sometimes difficult elsewhere in the country. Wind farms are usually located many miles from major electricity grids, and it is expensive to construct the necessary transmission lines (although the cost is very small compared with the cost of building a nuclear power plant!). Other problems currently bedevil these wind farmers. In North Dakota, most of the grid capacity ranging from Minneapolis-St. Paul to Chicago is dominated by coal burners, and farmers have had great difficulty gaining grid access, a problem that has generated severe political furor.[30]

However, there is good news for wind farmers. The John Deere Corporation and other companies that sell equipment to farmers are establishing large capital funds to invest in wind, supporting farmers and the production of renewable clean energy. Small local banks attracted to these potentially lucrative schemes are also beginning to invest money in wind power.[31]

It is imperative that the federal and state governments subsidize these important and critical new energy sources. Some states are already offering worthwhile subsidies. For instance, the Minnesota state Legislature is currently providing a production incentive to small wind farms, and the Minnesota Public Utilities Commission will purchase another 400 megawatts of wind capacity, having decided that wind power is the "least-cost alternative" for new electricity generation in the state.[32]

Farmers are also investigating other forms of green power including ethanol and soy diesel refineries in southwest Minnesota and anaerobic digesters that convert manure to green electricity. In short, with these types of alternatives, money that normally would be paid out to huge energy monopolies stays in the community, local jobs are created, local banks become involved, and communities prosper.[33]

Other countries such as China, with its hugely growing energy needs, have also begun to invest in wind power. In Huitengxile, on the grasslands of Inner Mongolia, a 68-megawatt wind farm has been established, which is expected to grow to 400 megawatts by 2008. Similar wind farms are being developed in many heavily populated provinces, and the cost per kilowatt of wind electricity is fast becoming competitive with China's abundant coal industry. Wang Zhongying, the director of China's Center for Renewable Energy Development, said that China has huge goals for wind power development, reaching 4,000 megawatts by 2010 and a staggering 20,000 megawatts by 2020.[34]

China supports the production of wind power and other alternatives with tax incentives for developers, while imposing standardized electricity rates as a subsidy for wind power, because it is still somewhat more expensive than coal. China has also ruled that provinces will be required to purchase electricity from alternative sources even if the cost per kilowatt hour is more expensive than conventional sources, a move that supports the suppliers of wind power.[35]

In England, wind farms are now providing megawatts of electricity to the national grid at a more rapid rate than those currently being lost as a result of nuclear power plant shut downs.[36]

Solar Power

Hypothetically 10 trillion to 20 trillion watts of solar power provided by photovoltaics could take the place of all conventional energy sources currently in use. Consequently, it has been estimated that a rather inefficient photovoltaic array covering half a sunny area measuring 100 square miles could meet all the annual U.S. electricity needs.[37] Although this is a vast amount of electricity, there are probably enough feedstocks—adequate and appropriate materials—to meet this gigantic challenge.

Photovoltaic cells are becoming both more efficient to produce and more efficient solar collectors. However, fossil-fuelled energy is necessary to create photovoltaic cells. A solar roof collector would therefore take one to four years to recover the amount of energy that produced it, but because it has a life expectancy of thirty years, 87% to 97% of the electricity it produces will not be plagued by pollution—greenhouse gases or resource depletion. There is ample space available to locate these solar arrays, including rooftops, alongside roadways, or on unused desert landcapes bathed in sun. The future production of massive numbers of solar collectors will require certain specialized materials, all of which are readily available, including even the rare minerals—indium and tellurium. The reliability, technological improvements, and market penetration of concentrated photovoltaics have all advanced considerably in the last twenty years.

Silicon Valley venture capitalists are now investing in "clean tech," a term encompassing solar energy, water purification systems, and alternative automotive fuels. These investors are not necessarily altruistic, although they recognize that doing good things for the planet is a "great by-product." (They freely admit that their motivation has a green tinge more to do with the color of money than with green energy.) Their investments are stimulated in part by the high price of oil but also by the increasing demand for electricity in India and China.[38]

Incentives encouraging solar power development are rapidly expanding and are now available in thirty states, including California, New York, and Texas, where in some cities the basic cost of conventionally generated electricity has risen 50% over the last three years. The legislators of these states recognize that they need to endorse creative technology to offset the large demand for electricity, without having to construct dirty new fossil-fuelled plants. Consequently, these solar initiatives are funded by alternative energy surcharges placed upon utility bills.[39]

Three-hundred thousand houses are now equipped with solar power, whereas five years ago the number was only 100,000. Sales of solar arrays increased 28% to $500 million during June 2004–2005 and this number is expected to rise more than 20% during June 2005–2006.[40] Because Governor Arnold Schwarzenegger is a solar enthusiast, solar installations have increased by 53% to power 4,614 residences in 2004 under California's rebate program. The Californian Public Utilities Commission in January 2006 passed the largest solar initiative in U.S. history approving a $3 billion rebate program to subsidize the installation of one million rooftop solar installations over the coming decade.[41] Other states are joining the solar fray. New Jersey offers the most generous rebates, whereas Connecticut and Ohio pay $5 per watt, Nevada offers $4 per watt, Oregon pays $3.50 per watt, and Idaho allows homeowners a 100% tax deduction up to $20,000 spent on the solar array.[42] Washington State has initiated a new solar tariff, similar to the one in Germany, to stimulate a major new demand in photovoltaics.[43] Forty states now allow homeowners to sell their excess solar power back to the grid.[44]

Despite these attractive and growing incentives, it is still relatively expensive to install a solar system. For example, a solar array installed in New Jersey cost the owner $50,000, of which the state provided $5.50 for each watt of generating capacity in his system, which covered $35,000 of the total cost. The owner is optimistic because he says he will recoup the balance of $15,000 in four to five years through savings on electricity bills and special credits earned through a state program akin to a frequent-flier mileage program to reward producers of non-polluting electricity.[45]

Moving on to other forms of renewable energy including solar power, Germany, which plans to phase out nuclear power by 2025, is moving rapidly toward alternatives. It now generates over 8% of its electricity from wind and biomass and is the world's largest user of photovoltaic cells. Because half its energy requirements will be generated from renewable sources by 2050, it predicts that carbon emissions will be reduced to one-fifth of its 1990 levels.[46]

In England, wave farms have moved from an experimental to a pragmatic stage and are now being heavily subsidized. This is where the fledgling wind power industry was five years ago (the first wave farm is now being launched in Portugal). Tidal power also has promising potential in Britain and may well become complementary to wind. At the same time, houses are being equipped with exciting new energy-saving gadgets.[47]

Some people question whether renewables could provide a practical electricity supply when wind power in various places can be intermittent and solar power changes according to season, climate, and the like.

Various studies have examined this "intermittency" problem related to renewables and solutions abound, including geographic aggregation of wind generators, improved weather forecasting techniques, timely extension of transmission and distribution grids, transboundary (between states and countries) of electricity exchange, and a mixture of renewable energy technologies including hydro, biomass, wind, solar, tidal, wave, geothermal, and cogeneration all interconnected on the same grid. This diversity will provide the full potential of renewables for adequate electricity production.[48]

Meanwhile, the U.S. Congress on October 6, 2005, allocated $50 billion more to the wars in Afghanistan and Iraq. One reason that the United States is involved in these countries is to control and own the oil. But oil burning adds to global warming. Would it not be better to withdraw from both these tortured countries and allocate these enormous sums of money to a world-class wind and solar economy? The money and technology are there, but the will and wisdom are not.[49]

10

What Individuals Can Do:
Energy Conservation and Efficiency

Europeans use approximately 50% less energy per capita than Americans, while maintaining the same standard of living.[1] Europeans are cognizant of energy use and conservation: a light turned on in a European hotel hallway is automatically extinguished within three minutes. Yet, with American advertising saturating global TV networks, the U.S. lifestyle has become the model for millions of people in China, India, Africa, and Indonesia, and even the Inuit in the Arctic.[2] If Americans change the way they live and decide to take responsibility to clean up the polluted planet, millions will follow.[3]

An aggressive U.S. agenda must be established by state and federal governments mandating reduced energy use at the *consumption* end of the equation, not just the production end. And individuals must urgently take responsibility for the way they live. While global warming is upon us, the earth is getting hotter; hurricanes and cyclones abound; and droughts, wildfires, and flooding rains destroy property, kill people, and devastate broad tracts of land.[4] Still, we drive SUVs, leave lights burning all over our houses, leave computers and VCRs on incessantly, and live at a constant temperature the whole year in heated and air-conditioned buildings.

Energy-efficient technologies have been available for many years, and they become more sophisticated daily.[5] Enough cost-effective energy-efficient measures and technologies are currently available to reduce electricity demand between 11% and 23% over the next five years and between 21% and 35% by 2020.

Aggressive, concerted, long-term public policy initiatives will be required to implement efficiency-related decisions in the market and to alter the way that people buy and use electricity appliances.[6] These decisions will be made only if state and federal governments take collective responsibility and conduct massive educational campaigns to inspire commercial enterprises and the public about the importance of conserving energy and how to do it.

Laws mandating responsible living must be enacted with urgency; for example, all new buildings should be constructed to be passive and active solar collectors, all extraneous lights must be extinguished at night in commercial buildings and in the home, and all electrical outlets that supply computers and other electronic equipment and all electrical fittings must be switched off at night except those that are absolutely necessary.

Home appliances are a very important facet of energy conservation. They are easy to install and use. Minimally, Americans and others should:

- use energy efficient light bulbs and light fixtures.
- use energy classified refrigerators and washing machines.
- use energy efficient dishwashers. (Or, better yet, discard your dishwasher and wash your dishes by hand in one sink full of water; that's what I do!)
- cease using clothes driers, which account for 6% of residential electricity use. Hang clothes out in the sun to dry in the summer and by the furnace in the winter.
- use solar hot-water systems. These are extremely efficient and readily available.
- lower the home temperature in the winter and turn off air-conditioning in the summer. In most but not all U.S. climates, air-conditioning is unnecessary for survival.

- completely weatherize houses with effective wall, ceiling, and floor insulation.
- build domestic dwellings to face south, acting as passive solar collectors.
- build overhanging eaves and plant deciduous trees to shade houses in the summer and provide sunlight in the winter.
- install a subsidized solar electricity generator on the roof.[7]

Commercial enterprises can use efficient lamps and ballasts (a part of an electrical circuit that regulates the current automatically under changing circumstances of voltage), adequate windows using sunlight instead of consistent artificial light, energy-efficient exit lamps and street and traffic lights, efficient heating and cooling systems, efficient office equipment, and energy management systems.[8] Industrial facilities should utilize efficient motors and motor drives, improved industrial processes, decent heating, ventilation, and cooling systems, efficient lighting and ballasts, and efficient energy management systems.[9]

Most energy-efficiency measures cost far less than the initial generation of electricity and its transmission and inefficient distribution over long distances. Other significant benefits accrue from their implementation as well.

- Energy-efficiency programs save huge amounts of money in statewide electricity costs and substantially reduce energy bills of customers.
- Energy efficiency significantly improves the environment, because for each kilowatt saved there is less electricity generation and less pollution from coal, oil, natural gas, or nuclear power.
- Energy efficiency promotes local economic development, creating jobs and increasing the disposable income of local citizens and businesses.

• Energy efficiency increases the independence of utilities
because they don't need to import fuels—coal, gas, oil, or
uranium—from other regions or countries. They therefore
need to generate less electricity, and they can become
smaller, more autonomous, and more efficient.[10]

Cities themselves are now instigating energy-saving initiatives.
The city of Portland, Oregon, is an outstanding example. Portland
managed to reduce CO_2 emissions below those of 1990 by encouraging a major increase in public transport, constructing 750
miles of new bicycle tracks and increasing the number of people
commuting by foot or bike by 10%. All city employees were offered a $25 per month bus pass or car pool parking, and city lights
have been replaced with energy-saving diodes. Financial incentives
and technical assistance are provided to people who construct
green buildings with built-in energy efficiency and who weatherize homes. The population is very proud and enthusiastic about
these developments because they are cheap, they reduce traffic con-
gestion, and the tax dollars saved on conservation are free to be
spent on more constructive activities. Mayor Tom Potter, who
drives a Prius hybrid, is excited and says that Portland is thriving
economically.[11]

CHOOSING A BETTER APPROACH TO ENERGY
IN THE FUTURE

A study performed in 2004 by Synapse Energy Economics, titled "A
Responsible Electricity Future," compared an analysis performed by
the U.S. Energy Information Administration under the Department
of Energy, which they call the "Reference Case" with Synapse's own
"Balanced Case" analysis. Where the government's Reference Case
extrapolated an increase in U.S. electricity consumption of more than

50% by 2025, the Balanced Case, if adequately applied, incorporates the use of energy efficiency, cogeneration, renewables, and natural gas, over a similar period of time to save an amount of electricity equivalent to that generated by more than 600 large, new power plants.[12]

The Synapse results are remarkable in their simplicity, ease of implementation, and positive effects upon climate change, whereas the Reference Case mandates enormous investment in expensive centralized fossil-fuelled generation, necessitating costly upgrades to the national electrical grid. Whereas the Reference Case forecasts that CO_2 emissions by 2025 will increase by 82% over 1990 CO_2 emissions, Balanced Case emissions will be 3% below 1990 emissions and 23% below 2000 emissions.[13] Furthermore, the Balanced Case would save $36 billion over twenty-four years.[14]

Here is Synapse's pragmatic plan to save the environment:

- *Energy efficiency* can reduce U.S. electricity demand by almost 28% by 2025 (compared with increased electricity use forecast by the Reference Case).
- Nonhydro *renewable energy*, including geothermal, landfill gas, biomass, solar thermal, solar power generation, and especially wind power, will provide 15% of U.S. electricity use by 2025 in the balanced case. (In the Reference Case, renewables account for only 1%, even less than the minuscule 2%, exclusive of hydropower, that these technologies currently account for.)[15]
- *Combined heat and power generation* will produce 10% (compared to 5% in the Reference Case).
- *Oil-, coal-, and gas-fired generators* are assumed to have been retired after fifty operating years. Some new coal plants will be incrementally constructed as the old ones are retired, but no new nuclear reactors will be built when all the old ones retire after forty-five years of operation.[16]

Cost

The Balanced Case not only saves $36 billion dollars but also considerably reduces the demands and constraints upon the overtaxed and inadequate U.S. transmission grid. First, by reducing the overall demand for electricity, it eases the demands on the grid. Second, by relying on cogeneration and renewable energy facilities, the Balanced Case advocates construction of smaller units situated near the end user, rather than massive nuclear power and coal plants, which are usually located hundreds of miles away.

The transmission and distribution (T&D) of centralized generated electricity is inefficient and very expensive. Typically, 10% of the electricity generated by large power plants is lost during transmission.[17] Currently the cost of T&D is $95 billion a year; by 2025 it will rise to $127 billion—a 34% increase according to the Reference Case. But in the Balanced Case, significantly lower costs will be incurred because the increase in load growth is so slow—a mere 4.7% by 2025—so the T&D costs are proportionally reduced.[18]

CO_2 Emissions

The reduction in CO_2 emissions in the Balanced Plan not only reduces global warming, but it will also lower the cost of electricity generation because there will be no need to comply with expensive regulations imposed upon energy corporations to minimize climate change.[19]

By comparison, physicist Dr. Tom Cochran extrapolated from the nuclear industry calculations for its future and found that by adding 700 gigawatts of nuclear electricity to the world—which would account for double today's nuclear capacity—for the fifty years from 2050 to 2100, this would entail:

- adding about 1,200 new nuclear plants—provided they last forty years and there are no meltdowns;
- adding fifteen new uranium enrichment plants;
- generating 0.97 million tons of high-level nuclear waste containing enough plutonium for hundreds of thousands of nuclear weapons;
- outfitting fourteen Yucca mountains to store the waste;
- adding fifty new reprocessing plants to extract plutonium if the Generation IV reactors were to proceed;
- investing $1 trillion to $2 trillion.

Far from decreasing gloabal warming, as the nuclear industry touts, the effect on the environment of this scenario would be to cut the global average temperature *rise* by just 0.2%.

Additional Benefits of the Balanced Case

Cost savings and benefits of the Balanced Case beyond those specifically enumerated include:

- lower carbon emission costs;
- lower economic and environmental costs associated with the decreased production of mercury, nitrous oxides, and particulate emissions;
- the benefits of decreased price volatility associated with the decreased use of fossil fuel;
- the environmental and health benefits accruing from decreased emissions, decreased land use for generators and for electricity transmission, and less water for power generation;
- environmental benefits associated with decreased mining for fossil fuel;
- increased jobs and economic benefits associated with renewable technologies.[20]

It is up to individuals throughout the world to choose a better approach to energy in the future. The nuclear option is neither desirable nor viable. But, as the Balanced Case makes clear, other options exist, and it is up to governments and citizens to implement them with urgency.

Global governments are under intense pressure to behave irresponsibly. Worried about the shortage of natural gas from Russia, the British government in January 2006 is considering proposals by British Nuclear Fuels to fast track the planning process by pre-licensing nuclear reactors before sites have even been selected, closing the process to public input.[21] However the Sustainable Development Commission in March 2006 concluded that a new British nuclear program would fail to answer the twin challenges of climate change and security of supply. It stated that a doubling of the United Kingdom's existing nuclear capacity would provide an 8% reduction in 1990 CO_2 levels, but zero CO_2 renewables would supply 68–87% of the United Kingdom's needs if fully exploited.[22]

I have found it very difficult to get Americans in particular to understand the urgency of the global warming situation, to comprehend the extraordinary dangers of nuclear power, and to develop the motivation and altruism to take personal and individual responsibility to protect the future for their children and descendents. There exists a deep sense of entitlement, a feeling that people can do anything and have anything they want, as long as they earn enough money.

But the world cannot be treated like that any longer. Global resources are finite and the misapplication of science and industry has seriously damaged the ecosystems of this unique planet, threatening the ongoing existence of many millions of species, including ourselves.

It is time, indeed beyond time, for Americans to step up to the true moral and spiritual responsibilities of belonging to the richest, most powerful nation on earth. American religious feeling, in its true sense—entailing responsibility for all creatures great and small—must be invoked. The democracy that Americans are so privileged and proud to share must be worked voraciously. Politicians, the people's representatives, must be educated and pursued with vigor to ensure that they do the right thing for the planet and not be manipulated and orchestrated by avaricious corporations.

Finally, living a moral life is a personal decision—to turn off the lights when you leave a room, to turn off the computer at night, to insulate the house, to wear more sweaters and turn down the heat in the winter, to practice using the sweat glands in the summer and not a global warming–air conditioner.

Self-sacrifice and responsibility are noble traits to which most people aspire. These are the qualities that will lead the world toward sanity and survival.

Notes

Introduction

1. http://www.whitehouse.gov/news/releases/2005/06/20050622.html, June 22, 2005.

2. "Nuclear Power and Children's Health, What You Can Do," a symposium presented by the Nuclear Policy Research Institute, the Nuclear Information and Resource Service, and Physicians for Social Responsibility–Chicago at St. Scholastica Academy, Chicago, October 15–16, 2004.

3. "Greenpeace Report Proves Solar Power Available to 100 Million People by 2025," Contact, Sven Teske, Greenpeace International Campaigner, Cairo, Egypt, 31621296.

4. "Waves Could Power 20% of the UK," Press Association, *The Guardian*, January 25, 2006.

5. Amory Lovins, "RMI's CEO Debunks Dangerous Nuclear Theology," *RMI Solutions*, http://www.rmi.org/images/other/Newsletter/NLRMIsummer 05.pdf, summer 2005.

6. Arjun Makhijani, "Our Electrical Future: A Non-Nuclear Low Carb Diet?" *New Hampshire Sierran*, Newsletter of the New Hampshire Sierra Club, Fall, 2005.

The production and use of electricity today is extremely inefficient. An electrical current consists of electrons, which pulse down a conducting wire. When an electric bulb is plugged into a socket, the electrons from the wire are converted to heat and to light. But the average efficiency of electric lighting systems is approximately 1%, meaning that only 1% of the energy in the fuel used to generate electricity produces visible light energy. All the rest is wasted as heat at the power plant or in the light bulb, as well as during transmission of the current through typically hundreds of miles of grid. Even high-efficiency light bulbs are only 3% efficient.

Conversely, electricity generated by a solar panel on a building is extremely efficient in part because there is no transmission loss.

7. Jonathan Leake and Dan Box, "When PR Goes Nuclear," *The Australian Financial Review*, May 27, 2005.

8. Ibid.

9. James Lovelock, *The Revenge of Gaia* (London: Penguin Books, 2006).

10. David Adam, "Next Generation of Nuclear Reactors May be Fast Tracked," *The Guardian*, January 21, 2006.

11. Ibid.

12. Ibid.

13. Laura Miller, "Nuclear Energy's Green Glow," *PR Watch*, May 24, 2005.

14. Felicity Barringer, "Old Foes Soften to New Reactors," *The New York Times*, May 15, 2005.

15. Miller, "Nuclear Energy's Green Glow."

16. Andrew Revkin, "Climate Expert Says NASA Tried to Silence Him," *The New York Times*, January 29, 2006.

17. Helen Thomas, "No Wonder Bush Doesn't Connect With the Rest of the Country," *Seattle Post Intelligencer*, October 15, 2003.

18. Stephen Schneider, "The Changing Climate," *Scientific American*, September 1989, 70–79.

19. John Nichols, "Enron: What Dick Cheney Knew," *The Nation*, April 15, 2005, 7–20.

20. Ibid.

21. Ibid.

22. Ibid.

23. Adam Klawoon, "Pro-Nuclear Conference Focuses on Tactics," *Union Tribune*, June 9, 2005.

24. "Nuclear Power—The Shape of Things to Come," *The Economist*, July 7, 2005.

25. Ibid.

26. Helen Caldicott, *Missile Envy* (New York: William Morrow, 2004).

1. The Energetic Costs of Nuclear Power

1. Lisa Rainwater van Suntrum, "Spinning Nuclear Power into Green," http://www.prwatch.org/prwissues/2005Q1/nuke2.html.

2. Personal communication during the 1980s with some Manhattan Project scientists, all of whom are now deceased and would not necessarily wish to be identified.

3. Some people are confused about the difference between global warming and ozone depletion. These are two totally different meteorological mechanisms. Ozone or O_3 accumulated in the upper layer of the atmosphere—the stratosphere—over billions of years, and this gas filters out the damaging carcinogenic ultraviolet light from the sun. The ozone has been thinned and even disappeared over some parts on the earth, such as the south pole, because a family of gases called chlorofluorocarbons, which were thought to be inert, were used widely in industry. These rose up into the stratosphere, combining with and destroying the ozone molecules. These gases linger in the stratosphere for 75 to 380 years. As the ozone layer declined, so the incidence of skin cancer among humans and animals rose, induced by an increased concentration of the sun's carcinogenic ultraviolet rays. For each 1% decrease in ozone, there is a 4% to 6% increase in skin cancer. In Australia, where the ozone layer is particularly thin, people are experiencing an epidemic of skin cancer and malignant melanoma. Because of this concern, CFC production was decreased under the Montreal Protocol in 1987, and many countries agreed to ban its production in 1990 in London. (Helen Caldicott, *If You Love This Planet*, New York: W.W. Norton, 1992). CFC gases are dangerous both because they destroy the ozone layer and because they are potent global warmers.

According to the U.S. Department of Energy (DOE) data, 93% of the CFC 114 gas released in the United States comes from uranium enrichment, the process by which fuel for nuclear power plants is created (Nancy Checklick, U.S. Department of Energy, e-mail message to author, June 8, 2004). Far from being "clean and green," this phase of nuclear energy production is by far the largest industrial emitter of a chemical that destroys the ozone layer (James Bruggers, "Uranium Plants Harm Ozone Layer, Kentucky, Ohio Facilities Top List of Polluters," *The Courier-Journal*, May 29, 2001).

4. Chlorofluorocarbon gases are potent global warmers, 10,000 to 20,000 times more effective than CO_2. Other global warming gases include methane or "natural gas," which is used to fuel gas-fired electricity plants, and nitric oxide, a component of car and coal power plant exhausts.

5. Jan Willem Storm van Leeuwen and Philip Smith, "Can Nuclear Power Provide Energy for the Future; Would it Solve the CO_2-emission Problem?" http://beheer.opvit.rug.nl/deenen/Nuclear_sustainabiliy_rev3.doc, October 12, 2004.

6. Ibid.

7. J.W. Storm van Leeuwen, "Nuclear Power—Some Facts," August 10, 2005, p. 10.

8. NEA-IAEA, *Uranium 2003: Resources, Production and Demand* (Paris: OECD, 2004).

9. A joule is a unit of energy; one joule is the energy of a heartbeat. A lamp of 100 watts consumes 100 joules of energy per second. To heat one liter of water one degree centigrade requires 4,800 joules or 4.8 kilojoules (KJ). A gigajoule is a billion joules.

10. Storm van Leeuwen and Smith, "Can Nuclear Power Provide Energy for the Future," Chapter 2, p. 4–8.

11. Ibid., Chapter 4, p. 4.

12. Ibid.

13. Personal communication with Brice Smith at IEER, Institute for Energy and Environmental Research, e-mail message to author, October 7, 2004.

14. Checklick, June 8, 2004.

15. Storm van Leeuwen and Smith, "Can Nuclear Power Provide Energy for the Future," Chapter 2, p. 9–10.

16. Ibid., Chapter 3, p. 10–12.

17. Ibid.

18. Ibid.

19. Ibid.

20. Ibid.

21. Ibid., Chapter 4, p. 2–8.

22. J.W. Storm van Leeuwen, "Radioactive Discharges from Nuclear Power: Sustainability and Nuclear Power," Chapter 5, November 16, 2005.

23. Storm van Leeuwen and Smith, "Can Nuclear Power Provide Energy for the Future," Chapter 4, p. 6.

24. Storm van Leeuwen, "Nuclear Power—Some Facts," p. 7.

25. BP Statistical Review of World Energy, June 2005, www.bp.com/statisti calreview2005; and Storm van Leeuwen and Smith, "Can Nuclear Power Provide Energy for the Future," Chapter 2, p. 12.

26. Storm van Leeuwen and Smith, "Can Nuclear Power Provide Energy for the Future," Chapter 5, p. 9.

27. The nuclear fuel cycle also uses large quantities of other deleterious compounds, many of which form potent global warming gases. To provide a single annual reactor fuel reload of 20.3 tons of enriched uranium 235, 162 tons of natural uranium must be converted to uranium hexafluoride and this process requires 77.6 tons of fluorine gas. The enriched 235 is then converted back to uranium oxide for nuclear fuel, releasing 9.72 tons of fluorine.

The remaining 141.7 tons of "depleted" uranium 238 hexafluoride gas (depleted of uranium 238) are stored in hundreds of thousands of steel containers in the open air. Because uranium hexafluoride gas is very reactive, almost certainly many of these steel drums are leaking. Worldwide, about 68,000 tons of natural uranium are fluoridated every year, requiring 32,600 tons of fluorine. Fluorine gas is dangerous and very reactive combining with many other chemical materials. It is therefore conceivable that very potent greenhouse gases are formed by fluorine reactions with organic solvents and that these gases are released as a by-product of uranium enrichment to the atmosphere. (J.W. Storm van Leeuwen, "Nuclear Power—the Energy Balance, Some Details of the Front End of the Nuclear Process Chain," November 18, 2005.)

Zirconium, the external cladding for the uranium fuel rods is purified with chlorine, potentially another source of potent global warming gases. The production of 7,600 to 15,200 tons of zircaloy annually for nuclear power plants requires a minimum of 11,700 to 23,400 tons of chlorine. Losses of chlorine are inevitable with significant amounts of chlorine compounds released to the atmosphere, and these chlorine gases may be responsible for the production of very potent global warming gases. (Storm van Leeuwen, "Nuclear Power—the Energy Balance," p. 2.)

As Storm van Leeuwen argues, "The nuclear industry should commit itself to publish a thorough analysis of the emissions of carbon dioxide and all other greenhouse gases in all processes of the fuel chain, before claiming that nuclear energy is carbon free or greenhouse gas free." (J.W. Storm van Leeuwen, "Uranium and Greenhouse Gases," August 13, 2005.)

28. Arjun Makhijani and Brice Smith, IEER, Institute for Energy and Environmental Research, July 26, 2005.

29. Ibid., Michael Mariotte, "Nuclear Power is Wrong Answer," *NIRS*, May 27, 2005; and Storm van Leeuwen and Smith, "Can Nuclear Power Provide Energy for the Future," Chapter 5, p. 7 -8

30. Storm van Leeuwen and Smith, "Can Nuclear Power Provide Energy for the Future?," Chapter 2, p. 12.

31. Personal e-mail communication with Jan Willem Storm van Leeuwen, March 11, 2006.

2. Paying for Nuclear Energy

1. Andrew Simms, Petra Kjell, and David Woodward, "Mirage and Oasis: Energy Choices in an Age of Global Warming," New Economics Foundation, http://www.neweconomics.org, June 29, 2005.

2. "Nuclear Price Tag," *New Scientist*, July 2, 2005.

3. "Nuclear Power the Shape of Things to Come? Climate Change Is Helping a Revival of the Nuclear Industry, Though Its Economics Still Look Dodgy," *The Economist*, July 7, 2005.

4. Ibid.

5. John Deutch, et al., "The Future of Nuclear Power: An Interdisciplinary MIT Study" (Cambridge, MA: Massachussetts Institute of Technology, 2003), p. 38.

6. "President Discusses Energy Policy, Economic Security," Calvert Cliffs Nuclear Power Plant, Lusby, Maryland, http://www.whitehouse.gov/infocus/energy, June 22, 2005.

7. Green Scissors, "Running on Empty: How Environmentally Harmful Energy Subsidies Siphon Billions from Taxpayers," Green Scissors campaign, http://www.foe.org/res/pubs/pdf/running.pdf, 2002 (cited February 18, 2005).

8. "NRDCs Perspective on Nuclear Power," NRDCs nuclear program, Natural Resources Defense Council Issue paper, June 2005.

9. Hermann Scheer, "Nuclear Energy Belongs in the Technology Museum," http://www.renewableenergyaccess.com/rea/news/story?id=19012. Renewable Energy Access (cited November 24, 2004).

10. Peter Bradford, "Nuclear Power's Prospects in the Power Markets of the 21st Century," Washington DC: The Non-Proliferation Education Center, January 2005.

11. "Nuclear Power—The Shape of Things to Come?"

12. Shankar Vedantam, "Uncertainties Slow Push for Nuclear Plants, Cost of Building New Facilities, Concerns About Waste Disposal Are Cited," Washington Post, July 24, 2005.

13. Ibid.

14. Christopher Sherry, "Throwing Good Money after Bad: Nuclear Power as a Clean Air 'Solution,'" Nuclear Information and Resource Service, http://www.nirs.org/climate/background/seccnukescleanair, July 19, 2005.

15. "Nuclear Power—The Shape of Things to Come?"

16. "Utilities Show Renewed Interest in Nuclear Power," AP, June 14, 2005.

17. Kelpie Wilson, "Exponential Enrons Ahead," Truthout/perspective, http://www.truthout.org/docs_2005/printer_062305A.shtml, June 23, 2005 (accessed June 24, 2005).

18. Ibid.

19. Ibid.

20. "Nuclear Power—The Shape of Things to Come?"

21. Philip Ward, "Unfair Aid: The Subsidies Keeping Nuclear Energy Afloat across the Globe," Nuclear Monitor, NIRS and WISE, #630–631, North American edition, June 30, 2005.

22. "U.S. Energy Legislation May Be 'Renaissance' for Nuclear Power," Bloomberg, June 22, 2005.

23. Ibid.

24. Ibid.

25. Natural Resources Defense Council, "NRDC's Perspective on Nuclear Power," NRDC nuclear program, Issue Paper, June 2005.

26. Ibid.

27. Bradford, "Nuclear Power's Prospects in the Power Markets of the 21st Century."

28. International Energy Agency, *Nuclear Power: Sustainability, Climate Change and Competition* (Paris: IEA, 1998).

29. Philip Ward, "Unfair Aid."

30. Personal e-mail with Frieda Berryhill, June 22, 2005.

31. Ibid.

32. European Environment Agency, "Energy Subsidies in the European Union: A Brief Overview," EEA Technical Report 1 (Copenhagen: EEA, 2004).

33. Ward, "Unfair Aid."

34. Ibid.

35. Nuclear Energy Agency, Organization for Economic Co-operation and Development, "Revised Nuclear Third Party Liability Conventions Improved Victims' Rights to Compensation," Press communiqué, http://www.nea.fr/html/general/press/2004/2004-01.html (cited February 13, 2004).

36. Ward, "Unfair Aid."

37. "Energy Supply and Other Defense Activities," Office of Nuclear Energy, Science and Technology, FY 2005, Congressional Budget.

38. Ibid.

39. M. Goldberg, "Federal Energy Subsidies: Not All Technologies Are Created Equal," REPP Research Report, http://www.crest.org/repp_pubs/pdf/subsidies.pdf, July 2000 (accessed April 23, 2005).

40. A. Froggatt, "The EU's Energy Support Programmes: Promoting Sustainability or Pollution?" Greenpeace International, http://greenpeace.org/international_cn/multimedia/download/1/459479/0/EUsubsidiesReport.pdf, April 2, 2004 (accessed March 1, 2005).

3. Nuclear Power, Radiation, and Disease

1. www.ncbi.nlm.nih.gov/omim/mimstats.htm/, March 8, 2006.

2. Richard R. Monson, Chair, and James S. Cleaver, Vice Chair, "Low Levels of Ionizing Radiation May Cause Harm," The National Academy of Sciences, BEIR VII report, June 29, 2005.

3. Ibid.

4. Ibid.

5. Ibid.

6. Susan S. Devesa, et al., "Recent Cancer Trends in the United States," *Journal National Cancer Institute* 87 (February 1995): 175–82.

7. J.G. Gurney, et al., "Trends in Cancer Incidence among Children in the U.S.," *Cancer* 78, no. 3 (August 1, 1996): 532–41; and J.J. Mangano, "A Rise in the Incidence of Childhood Cancer in the United States," *International Journal of Health Services* 29, no. 2 (1999): 393–408.

8. Monson and Cleaver, "Low Levels of Ionizing Radiation May Cause Harm."

9. Ibid.

10. Jon D. Erickson, Duane Chapman, and Ronald E. Johnny, "Monitored Retrievable Storage of Spent Nuclear Fuel in Indian Land: Liability, Sovereignty, and Socioeconomics," *American Indian Law Review* 19, no. 1 (1994): 88.

11. "Navajos 'Chop the Legs Off the Uranium Monster,'" *WISE/NIRS Nuclear Monitor*, May 13, 2005.

12. Keith Schneider, "A Valley of Death for Navajo Uranium Miners," *New York Times*, May 3, 1993; and Patricia Kahn, "A Grisly Archive of Key Cancer Data," *Science* 259 (January 22, 1993).

13. "Navajos 'Chop the Legs Off the Uranium Monster.'"

14. Ibid.

15. Ibid.

16. Michael E. Long, "Half Life: The Lethal Legacy of America's Nuclear Waste," *National Geographic*, July 2002.

17. Helen Caldicott, *Nuclear Madness* (New York: Norton, 1994); and Erickson, Chapman, and Johnny, "Monitored Retrievable Storage of Spent Nuclear Fuel in Indian Land."

18. Ibid.

19. Caldicott, *Nuclear Madness*.

20. Long, "Half Life."

21. Personal e-mail from Doug Rokke, March 11, 2006, former Director of U.S. Army Depleted Uranium Project.

22. Caldicott, *Nuclear Madness*.

23. Ibid.

24. J.W. Storm van Leeuwen, "Radioactive Discharges from Nuclear Power," in *Sustainability and Nuclear Power*, November 16, 2005.

25. E-mail communication with David Lochbaum, Union of Concerned Scientists, September 2005.

26. T. Chandrasekaran, J.Y. Lee, and Charles A. Willis, "Calculation of Re-leases of Radioactive Materials in Gaseous and Liquid Effluents from Pressurized Water Reactors (PWR–GALE code)," U.S. Nuclear Regulatory Commission, Office of Nuclear Reactor Regulation-0017, April 1978 (the most recent version).

27. Caldicott, *Nuclear Madness*.

28. Steve Wing, "Objectivity and Ethics in Radiation Biology," *Environmental Health Perspectives* 3, no. 14 (November 2003).

29. Kay Drey, "Why Routine Radioactive Releases Occur," unpublished paper, University City, MO: 1985.

30. Ibid.

31. Chandrasekaran, Lee, and Willis, "Calculation of Releases of Radioactive Materials."

32. Ibid.; R. Lowry Dobson, "The Toxicity of Tritium," International Atomic Energy Commission Symposium, *Biological Implications of Radionuclides Released from Nuclear Industries*, 1 (Vienna: IAEA, 1979): 203; P. Torok, W. Schmahl, I. Meyer, and G. Kistner, "Effects of a Single Injection of Tritiated Water during Organogeny on the Prenatal and Postnatal Development of Mice," International Atomic Energy Commission Symposium, *Biological Implications of Radionuclides Released From Nuclear Industries*, 1 (Vienna: IAEA, 1979): 241; and T.E.F. Carr and J. Nolan, "Testis Mass Loss in the Mouse Induced by Tritiated Thymidine, Tritiated Water and 60Co Gamma Radiation," *Health Physics* 36 (February 1978): 135–145.

33. T. Rytomaa, J. Saltevo, and H. Toivonen, "Radiotoxicity of Tritium Labelled Molecules," International Atomic Energy Commission Symposium, *Biological Implications of Radionuclides Released From Nuclear Industries*, 1 (Vienna: IAEA, 1979): 339; and Z. Pietrzak-Flis, I. Radwan, A. Major, and M. Kowalska, "Tritium Incorporated in Rats Chronically Exposed to Tritiated Food or Tritiated Water for Three Successive Generations," *Journal of Radiation Research*, 22 (1982): 434–42.

34. DOE—HDBK—1079-94 Primer on Tritium.

35. J.R. Watts and C.E. Murphy Jr., "Assessment of Potential Radiation Dose to Man From an Acute Tritium Release into a Forest Ecosystem," *Health Physics* 35 (August): 287–91.

36. Drey, "Why Routine Radioactive Releases Occur."

37. Ibid.

38. Dr. Ernest Sternglass, e-mail to the author, August 2, 2005.

39. Ibid.

40. Ibid.

41. Kay Drey, telephone conversation, October 2005.

42. Chandrasekaran, Lee, and Willis, "Calculation of Releases of Radioactive Materials."

43. I first learned about the medical effects of nuclear power in 1975 when I read a book called *Poisoned Power* by Gofman and Tamplin. Ever since then I have been asking the nuclear industry what they are going to do with their waste. They tell me in a patronizing tone of voice that they are good scientists and that one day they will find a solution to the confinement of their radioactive sewage. Never have I received a satisfactory answer to my question. This is akin to my telling a patient that he has a pancreatic cancer and will probably die within six months, but reassuring him by saying that I am a good scientist, and that in twenty years time I will have found the cure.

44. Grigori Medvedev, *The Truth about Chernobyl* (New York: Basic Books, 1991).

45. Brookhaven National Laboratory Annual Report to EPA Region II for 1993 Compliance with 40 C.F.R. 61, Section 94 Reporting Requirements, table 2: Facility Radionuclide Emissions, "Removal of the BGRR Above Ground Ducts Projects," June 8, 2000, Revision 2, p. 6. Surface contamination of ducts of CS137 ranges from 8,500 ± 2600 pci/cm² to 24,100 ± 1600 pci/cm² (pci = picocuries).

45a. Suffolk County Law Resolution No. 728-2000, amended by Resolution No. 906-2000 "Creating the Suffolk County Legislature Rhabdo-myoma Task Force."

46. John W. Gofman, *Radiation-Induced Cancer from Low-Dose Exposure: An Independent Analysis*, 1st ed. (San Francisco: Committee for Nuclear Responsibility, 1990), p. 16.

47. Greg Minor, former G.E. nuclear engineer, MHB Technical Associates, telephone conversation, November 1992; and Daniel F. Ford, Henry W. Kendall, and Lawrence S. Tye, Union of Concerned Scientists, "Browns Ferry and Regulatory Failure," June 10, 1976.

48. Wing, "Objectivity and Ethics in Environmental Health Science."

49. Steve Wing, telephone conversation, November 2005.

50. Caldicott, *Nuclear Madness*.

51. Dr. Karl G. Morgan, "Missing and Inadequate Data on Radionuclide Releases and Population Doses Resulting from TMI-2 Accident of March 28, 1979—Reasons for Concern," notes made on March 24, 1982.

52. M. Wahlen, et al., "Radioactive Plume from the Three Mile Island Accident: Xenon-133 in Air at a Distance of 375 Kilometers," *Science* 297 (February 8, 1980).

53. Personal telephone conversation with David Lochbaum on January 21, 2005.

54. "The American Experience: Meltdown at Three Mile Island," PBS, March 1999.

55. Morgan, "Missing and Inadequate Data on Radionuclide Releases and Population Doses Resulting from TMI-2 Accident of March 28, 1979."

56. Carl J. Johnson, MD, "Transuranics and the Impact on Health," statement made at press conference, Washington DC, May 28, 1985.

57. Paul Gunter, director of the Reactor Watchdog Project, Nuclear Information Resource Services, telephone conversation, November 10, 2005. Three Mile Island is not the only reactor to empty its waste water and cooling water into Chesapeake Bay. The ten other reactors draining into the bay include Susquehanna 1 and 2, Peach Bottom 2 and 3, Calvert Cliffs 1 and 2, North Anna 1 and 2, and Surry 1 and 2. Each reactor uses approximately 1 million gallons of cooling water a minute, which it then discharges as relatively radioactive water into the river.

58. "Suit Claims Damage from TMI Venting," *The Daily News*, June 23, 1982.

59. Beth Snyder, "TMI to Dispose of Contaminated Water," *York Daily Record*, November 28, 1990.

60. Wing, "Objectivity and Ethics in Environmental Health Science."

61. John C. Stauber and Sheldon Rampton, "Spin Dr. Strangelove, or How We Learned to Love the Bomb," *PR Watch*, 2, no. 4, Fourth Quarter, 1995.

62. Caldicott, *Nuclear Madness*.

63. Wing, "Objectivity and Ethics in Environmental Health Science."

64. Joseph Mangano, "Three Mile Island: Health Study Meltdown," *Bulletin of the Atomic Scientists* (September/October 2004).

65. Food and Drug Administration Sample Analysis, Three Mile Island Accident, April 4, 1979.

66. Interoffice memorandum on Hershey's stationery, C.J. Crowell to W.J. Crook, April 11, 1979.

67. Letter to Dr. Carl Y. Wong, Group Leader Product Research, Hershey Foods Corporation Research Laboratories, from K.K.S. Pillay, Associated Professor, Nuclear Engineering, Pennsylvania State University, College of Engineering, April 16, 1979.

68. Richard Kauffman, "Iodine 131 Levels Said Insignificant," *Sunday Patriot News*, Harrisburg PA, April 8, 1979.

69. Wing, "Objectivity and Ethics in Science."

70. Ibid.

71. Ibid.

72. Ibid.

73. Ibid.

74. Ibid.

75. Ibid.

76. Mangano, "Three Mile Island: Health Study Meltdown."

77. Karl Z. Morgan, "Health Physics: Its Development, Successes, Failures and Eccentricities," *American Journal of Industrial Medicine* 22 (1992): 125–33.

78. Telephone conversation with Greg Minor, 1994.

79. "EFMR Completes Settlement Requirements," 2005 biennial report, EFMR Monitoring Group, http://www.efmr.org (accessed November 10, 2005).

80. Marian Uhlman, "Study Shows High Cancer Rates in Area," *Philadelphia Enquirer*, November 19, 2002.

81. Mary Warner, "$3.9 million OK'd for TMI Injury Claims," *The (Harrisburg, PA) Patriot*, February 7, 1995.

82. Mycle Schneider, "The Chernobyl Disaster: A Human Tragedy for Generations to Come," *IPPNW Global Health Watch*, no. 4 (Cambridge, MA: IPPNW, September 2004); Chernobyl Forum Report, 2005, http://www.iaea.org/Newscenter/Focus/Chernobyl/pdfs/05-28601_Chernobyl. pdf.

83. Richard Bramhall, Chris Busby, and Paul Dorfman, *CERRIE Minority Report 2004*, UK Department of Health/Department of Environment Committee Examining Radiation Risks of Internal Emitters, Aberystwyth: Sosiumi Press, 2004.

84. Medvedev, *The Truth about Chernobyl*, p. 32.

85. Gofman, *Radiation-Induced Cancer from Low-Dose Exposure.*

86. Schneider, "The Chernobyl Disaster."

87. Julie Godoy, "French Finally Confront Chernobyl Risks," *IPS*, April 1, 2005.

88. Ibid.

89. Schneider, "The Chernobyl Disaster."

90. Godoy, "French Finally Confront Chernobyl Risks."

91. Report of the Government of Ukraine, Annex III of UNSG, "Optimizing the International Effort to Study, Mitigate and Minimize the Consequences of the Chernobyl Disaster," Report of the Secretary General, UN General Assembly, August 29, 2003; In 1991 the Ukranian government noted that 2,000 people had "disabilities connected with the Chernobyl disaster," but this number increased to almost 100,000 by January 1, 2003; In 2000 the World Health Organization predicted that 500,000 new cases of thyroid cancer will arise among young people living in the worst affected regions.

The New Scientist reported that there is a 90-fold increase in thyroid cancer in the most contaminated regions, and that there is a strong correlation between radiation dose and the incidence of cancer.

92. Schneider, "The Chernobyl Disaster."

93. UN-OCHA, "Chernobyl: Needs Great 18 Years after Nuclear Accident," United Nations Office for the Coordination of Human Affairs, press release, New York, April 26, 2004.

94. UNDP, UNICEF, "The Human Consequences of the Chernobyl Nuclear Accident—A Strategy for Recovery," Report commissioned by UNDP and UNICEF with the support of UN-OCHA and WHO, January 25, 2002.

95. Martin Tondel, et al., "Increase of Regional Total Cancer Incidence in North Sweden Due to the Chernobyl Accident?" *Journal of Epidemiology and Community Health* 58 (2004): 1011–16.

96. Godoy, "French Finally Confront Chernobyl Risks."

97. Dr. David R. Marples, "Chernobyl Ten Years Later—The Facts," University of Alberta, Canada, March 21, 1996.

4. Accidental and Terrorist-Induced Nuclear Meltdowns

1. Paul Gunter, Director of the Reactor Watchdog Project, Nuclear Information Resource Services, telephone conversation, November 10, 2005.

2. Paul Gunter, Director of the Reactor Watchdog Project, "Davis-Besse Nuclear Plant Comes Close to Disaster as Lax Regulator Places Company Interests Ahead of Public Safety," Nuclear Information Resource Services, March 13, 2002.

3. David Lochbaum, *US Nuclear Plants in the 21st Century: The Risk of a Lifetime* (Cambridge, MA: Union of Concerned Scientists, 2004).

4. Ibid.

5. *NRC Information Digest*, NUREG 1350, 16, rev. 1, 2004–2005.

6. David Lochbaum, Testimony on Nuclear Power before the Clean Air, Wetlands, Private Property, and Nuclear Safety Subcommittee of the United States Senate Committee on Environment and Pubic Works, May 8, 2001. Union of Concerned Scientists, 1707 H Street, NW, Suite 600, Washington DC, 20006-3919.

7. David Lochbaum, "Nuclear Power Plant Safety in Region C," in *US Nuclear Power Plants in the 21st Century: The Risk of a Lifetime* (Cambridge, MA: Union of Concerned Scientists, 2004).

8. Ibid.

9. Ibid; and report from state of New York, February 15, 2000, "Event at Indian Point 2," http://www.dps.state.ny.us/ip2report.htm.

10. Ibid; Letter from Steven Long, NRC, to John Groth, Senior Vice President, Nuclear Operations, Consolidated Edison Company of New York, November 20, 2000—subject: "Final Significance Determination for a Real Finding and Notice of Violation at Indian Point 2"; and letter to Honorable Edward J. Markey, August 17, 2004 from the Nuclear Regulatory Commission.

11. Ibid.

12. Lochbaum, Testimony on Nuclear Power before the Clean Air, Wetlands, Private Property, and Nuclear Safety to the Subcommittee on Environment and Public Works.

13. Ibid.

14. Ibid.

15. Lochbaum, "Nuclear Power Plant Safety in Region C."

16. Ibid.

17. Paul Schwartz, "For Nuclear Power, the Heat Is On," WBAI Pacifica Radio, November 11, 2003.

18. Wolf Blitzer, CNN Anchor, August 12, 2003; and "France Frets Over Nuke Plants and Heatwave Toll," Planet Ark, France, August 12, 2003.

19. Ibid.

20. Keay Davidson, "Reassessing 'What If' Factor at State's Nuclear Power Plants: December Tsunami Prompts Scientists to Review All Risks," San Francisco Chronicle, July 11, 2005; and Humbolt Bay, 1.0 Site Identification, Location Eureku, CA, Licence No. DPR-7, Docket No: 50-133, Project Manager, John Hackman, September 16, 2005.

21. Edwin S. Lyman, "Chernobyl on the Hudson: The Health and Economic Impacts of a Terrorist Attack at the Indian Point Nuclear Plant," Union of Concerned Scientists, September 2004.

22. Ibid.

23. John H. Large, "The Aftermath of September 11: The Vulnerability of Nuclear Plants to Terrorist Attacks," IPPNW Global Health Watch, Report no. 4, 2004.

24. Ibid.

25. Daniel Hirsch, David Lochbaum, and Edwin Lyman, "The NRC's Dirty Little Secret," Bulletin of the Atomic Scientists, May/June 2003.

26. Large, "The Aftermath of September 11."

27. Ibid.

28. Hirsch, Lochbaum, and Lyman, "The NRC's Dirty Little Secret."

29. Ibid.

30. Mark Thompson, "Are These Towers Safe? Why America's Nuclear

Power Plants Are Still So Vulnerable to Terrorist Attack—and How to Make Them Safer," *Time*, June 20, 2005.

31. Ibid.

32. Ibid.

33. Hirsch, Lochbaum, and Lyman, "The NRC's Dirty Little Secret."

34. Ibid.

35. Ibid.

36. Thompson, "Are These Towers Safe?"

37. Ibid.

38. Ibid.

39. Ibid.

40. Ibid.

41. Ibid.

42. The data in this section are drawn from an excellent study of a meltdown at Indian Point published by Dr. Ed Lyman from the Union of Concerned Scientists. Edwin S. Lyman, "Chernobyl on the Hudson: The Health and Economic Impacts of a Terrorist Attack at the Indian Point Nuclear Power Plant," Union of Concerned Scientists, September 2004.

43. Ibid.

44. Ibid.

45. Ibid.

46. Ibid.

47. Ibid.

48. Ibid.

49. Ibid.

50. Ibid.

51. Ibid.

52. Ibid.

53. Ibid.

54. Ibid. Lyman's report does find that for the particular scenario and sheltering parameters used in the model, sheltering makes sense. But the report also points out that these results are very scenario-dependent and does not go as far as calling for sheltering to replace evacuation as the preferred strategy.

55. Ibid.

56. Ibid.

57. Ibid.

58. Ibid.

59. The NRC in 1998 produced a most damaging report called NUREG-

1633 in which they stated that adult thyroid doses within the ten-mile range could vary between 1,500 to 19,000 rems during a severe weather sequence—rain. They stated that such thyroid doses are definitely possible during severe nuclear accidents. (F. J. Congel, et al. "Assessment of the Use of Potassium Iodide (KI) as a Public Protective Action During Severe Reactor Accidents," Draft report for comment, NUREG-1633, U.S. Nuclear Regulatory Commission, July 1998.) However, the NRC has mysteriously withdrawn this important and damaging document from public access (Lyman, "Chernobyl on the Hudson").

60. Lyman, "Chernobyl on the Hudson."

61. World Health Organization, *Guidelines for Iodine Prophylaxis Following Nuclear Accidents* sec. 3.3 (Geneva: WHO, 1999).

62. Ibid.

63. Lyman, "Chernobyl on the Hudson."

64. Ibid.

65. Ibid.

66. Information for the remainder of this chapter has been gleaned from a paper published in *Sciences and Global Security* in 2003. Robert Alveraz, et al., "Reducing the Hazards from Stored Spent Power Reactor Fuel in the United States," *Science and Global Security* 11 (2003): 1–51.

67. Personal e-mail communication with David Lochbaum, March 15, 2006.

68. Robert Alveraz, et al., "Reducing the Hazards from Stored Spent Power Reactor Fuel in the United States," *Science and Global Security,* 11:1–51, 2003.

69. Geoffrey Lean, "Attack on Nuclear Plant 'Could Kill 3.5 m,'" The Independent, Online Edition, http:/news.independent.co.uk/uk/environment/story .jsp?story=378739, February 16, 2003.

5. Yucca Mountain and the Nuclear Waste Disaster

1. Other kinds of nuclear waste include transuranic waste, low- and mixed low-level waste, and tailings. Transuranic waste is material contaminated with plutonium or its deadly alpha emitting relatives—neptunium, americium, curium, einsteinium and others—consisting of tools, clothing, filters, and other polluted matter. A salt mine been opened in Carlsbad, New Mexico, to receive some of this transuranic waste, but 11.3 million tons remain buried at numerous government sites.

Low- and mixed low-level waste includes hospital, industrial, research, and institutional waste and polluted materials from air filters, clothing, decommissioned power plants, and tools, which amounts to approximately 472 million cubic feet. Tailings from uranium mining and milling amount to 265 million tons.

Time magazine encompassed this waste problem in an article published in July 2002 written by Peter Essick when he wrote, "Load these tailings into railroad cars, then pour the 91 million gallons of waste into tank cars, and you would have a mythical train that would reach around the equator and then some" (Peter Essick, "Half Life NRC, NUREG-1350, The Lethal Legacy of America's Nuclear Waste," *Time*, http://www.nrc.gov/reading-rm/doc-collections/nuregs/staff/sr1350, July 2002).

2. Dr. Paul P. Craig, former member of the U.S. Nuclear Waste Technical Review Board, "Yucca Mountain—Time to Slow Down," August 14, 2005.

3. Ibid.

4. "Background Status of High-Level Nuclear Waste Management," Nuclear Information and Resource Service, August 1992; "Stop the Yucca Mountain Nuclear Waste Dump," Greenpeace, c/o Nuclear Information and Resource Service; and personal e-mail communication with Paul Craig, January 17, 2005.

5. Personal e-mail communication with Paul Craig, January 17, 2005.

6. Presentation to the California Energy Commission, energy.ca.gov/ 2005_energypolicy/documents/2005-08-15–16_workshop/presentations/panel-1/Craig_Yucca_Mountain.pdf, August 14, 2005.

7. Helen Caldicott, *Nuclear Madness* (New York City: WW Norton, 1994).

8. Personal e-mail communication with Judy Treichel, Executive Director, Nevada Nuclear Waste Task Force, January 22, 2006.

9. Direct quote from Paul Craig, used by Treichel.

10. Caldicott, *Nuclear Madness*; and personal e-mail communication with Treichel.

11. Nevada State web site, "Chronology of Selected Yucca Mountain Emails," http://www.state.nv.us/nucwaste/September 9, 2007.

12. Ibid.

13. Ibid.

14. Ibid.

15. Ibid.

16. "EPA Proposing Radiation Exposure Limits," AP, August 9, 2005.

17. Craig, "Yucca Mountain."

18. "EPA Proposing Radiation Exposure Limits," AP, August 9, 2005.

19. Nevada State web site, "Chronology of Selected Yucca Mountain Emails."

20. http://www.state.nv.us/nucwaste/news2005/pdf/ymchron0.1.pdf.

21. Personal e-mail communication with Craig.

22. Personal telephone conversation with Paul Gunter, Director of the Reactor Watchdog Project, Nuclear Information Research Services, March 14, 2006.

23. "EPA Proposing Radiation Exposure Limits," Associated Press, August 10, 2005.

6. Generation IV Nuclear Reactors

1. American Nuclear Society, "World List of Nuclear Power Plants," *Nuclear News*, March 2005.

2. Ibid.

3. Helmut Hirsch, Oda Becker, Mycle Schneider, Anthony Froggatt, "Nuclear Reactor Hazards: Ongoing Dangers of Operating Nuclear Technology in the 21st Century," Greenpeace International, April 2005; and David Lochbaum, "A Plan for Change or a Worst-Case Scenario," February 8, 2005.

4. American Nuclear Society, "World List of Nuclear Power Plants."

5. Anthony DePalmer, "Canadians Export a Type of Reactor They Close Down," *New York Times*, December 3, 1997.

6. Hirsch, et al., "Nuclear Reactor Hazards, Ongoing Dangers of Operating Nuclear Technology in the 21st Century."

7. Ibid.

8. "Magnox Reactors," http://www.westinghousenuclear.com/C1a8.asp.

9. Hirsch, et al., "Nuclear Reactor Hazards."

10. Ibid.

11. Ibid.

12. Ibid.

13. David Lochbaum, statement submitted to the House Government Reform Subcommittee on Energy Resources, "The Next Generation of Nuclear Power," June 29, 2005.

14. Ibid.

15. David Brewer, "Nuke Plant Plans Aired," *Huntsville (AL) Times*, August 5, 2005.

16. Hirsch, et al., "Nuclear Reactor Hazards."

17. Ibid; and Lochbaum, "The Next Generation of Nuclear Power."

18. Ibid.

19. "Advanced Reactor Study," MHB Technical Associates, Consultants on Energy, released by the Union of Concerned Scientists, July, 1980.

20. Hirsch, et al., "Nuclear Reactor Hazards."

21. "Advanced Reactor Study."

22. Personal e-mail communication with David Lochbaum, January 9, 2006.

23. Ibid.

24. David Lochbaum, testimony on nuclear power before the Clean Air, Wet-

lands, Private Property, and Nuclear Safety Subcommittee of the United States Senate Committee on Environment and Public Works, May 8, 2001.

25. John Deutch, et al., "The Future of Nuclear Power: An Interdisciplinary MIT Study," p. 38, Cambridge, MA: Massachussetts Institute of Technology, 2003.

26. "NRDC's Perspective on Nuclear Power," Natural Resources Defense Council, Issue Paper, June 2005.

27. Personal telephone conversation with David Lochbaum, March 11, 2006.

28. "Windscale Fire," Wikipedia, http://en.wikipedia.org/wiki/wind scale_fire.

29. "NRDC's Perspective on Nuclear Power."

30. Ibid.

31. Personal telephone communication with Tom Cochran, NRDC, September 2005.

32. Hirsch, et al., "Nuclear Reactor Hazards"; and "Advanced Reactor Study."

33. Hirsch, et al., "Nuclear Reactor Hazards."

34. Lochbaum, "A Plan for Change or a Worst-Case Scenario."

35. Personal telephone conversation with Lochbaum, March 11, 2006; and statement submitted by David Lochbaum to the House Government Reform Subcommittee on Energy Resources.

36. Statement submitted by David Lochbaum to the House Government Reform Subcommittee on Energy Resources.

37. Hirsch, et al., "Nuclear Reactor Hazards."

38. Ibid.

39. Ibid.

40. Ibid.

41. Ibid.

7. Nuclear Energy and Nuclear Weapons Proliferation

1. Ian Cobain and Ewen MacAskill, "Nuclear Arms Supermarket Doing a Roaring Trade," *Sydney Morning Herald*, October 10, 2005.

2. "NRDC's Perspective on Nuclear Power," Natural Resources Defense Council, Issue Paper, June 2005.

3. Allison MacFarlane, Frank von Hippel, Jungmin Kang, and Robert Nelson, "Plutonium Disposal: The Third Way," *Bulletin of the Atomic Scientists*, May/June 2001.

4. Helen Caldicott, *Nuclear Madness* (New York: W.W. Norton, 1994).

5. MacFarlane, et al., "Plutonium Disposal the Third Way."

6. Ibid.

7. Arjun Makhijani, "Nuclear Power, No Solution to Global Climate Change," Paper, Takoma Park, MD: Institute for Energy and Environmental Research, March 1998.

8. Ibid.

9. Personal e-mail communication with David Lochbaum, Union of Concerned Scientists, January 18, 2005.

10. Ibid.

11. "NRDC's Perspective on Nuclear Power," Natural Resources Defense Council, Issue Paper, June 2005.

12. "Uranium Enrichment and Fuel Fabrication—Current Issues," WISE Uranium Project, http://www.wise-uranium.org (accessed September 6, 2005).

13. Ibid.

14. Zia Mian and M.V. Ramana, "Feeding the Nuclear Fire," *Foreign Policy in Focus* (September 20, 2005); and "NRDC's Perspective on Nuclear Power."

15. James Sterngold, "Experts Fear Nuke Genie's Out of the Bottle: Arms Technology Spreading Beyond Iran, North Korea," *San Francisco Chronicle*, November 22, 2004.

16. Ibid.

17. Ibid.

18. Mian and Ramana, "Feeding the Nuclear Fire."

19. Helen Caldicott, *The New Nuclear Danger: George Bush's Military Industrial Complex*, New York: The New Press, 2004.

20. Bruce G. Blair, Harold A. Feiveson, and Frank N. von Hippel, "Taking Nuclear Weapons Off Hair Trigger Alert," *Scientific American* (November 1999).

21. "First Irradiated Tritium Rods Arrive at SRS, NNSA Readiness Campaign Reaches Milestone," National Nuclear Security Administration, September 9, 2005.

22. Walter Pincus, "Pentagon Revises Nuclear Strike Plan; Strategy Includes Preemptive Use against Banned Weapons," *Washington Post*, September 11, 2005.

23. "Draft US Defense Paper Outlines Preventive Nuclear Strikes," *Agence France Presse,* September 11, 2005.

24. "Plan Envisions Using Nukes on Terrorists," AP, September 11, 2005.

25. "Bush Administration's Nuclear Policy Slammed by Einstein Associate: Nuclear Arsenals Create More Danger Now than during the Cold War," Press Release Newswire, http://www.prweb.com/releases/2004/10/prweb172783.htm, October 28, 2004.

8. Nuclear Power and "Rogue Nations"

1. Nazila Fathi, "Defending Nuclear Ambitions, Iranian President Attacks US," *New York Times*, November 27, 2005.

2. Larry Rohter and Juan Forero, "Venezuela's Leader Covets a Nuclear Energy Program," *New York Times*, November 27, 2005.

3. Tony Benn, "Bush Is the Real Threat," *The Guardian*, August 31, 2005.

4. Pierre Goldschmidt, "Decision Time on Iran," *New York Times*, September 14, 2005.

5. "Russia Wants to Build More Nuke Reactors for Iran," Reuters, June 28, 2005.

6. Mark Landler, UN Says It Hasn't Found Much New about Nuclear Iran," *New York Times*, September 3, 2005.

7. "Iran Gives IAEA Bomb Part Instructions," Reuters, November 18, 2005.

8. "Russia Wants to Build More Nuke Reactors for Iran," Reuters, June 28, 2005.

9. Mark Landler, "Nuclear Agency Votes to Report Iran to UN," *New York Times*, September 25, 2005.

10. Goldschmidt, "Decision Time on Iran."

11. Seymour M. Hersh, "The Iran Plans," *The New Yorker,* April 17, 2006.

12. Michael T. Klare, "The Iran War Buildup," *The Nation*, http://www.thenation.com, July 21, 2005.

13. Ibid.; Seymour M. Hersh, "The Coming Wars," *The New Yorker*, January 24 and 31, 2004.

14. Klare, "The Iran War Buildup."

15. Ibid.

16. William Arkin, "Not Just a Last Resort? A Global Strike Plan, with a Nuclear Option," *Washington Post*, May 15, 2005.

17. Klare, "The Iran War Buildup."

18. Joel Brinkley, "Iranian Leader Refuses to End Nuclear Effort," *New York Times*, September 18, 2005.

19. Ibid.

20. Ibid.

21. Helen Caldicott, *The New Nuclear Danger: George Bush's Military Industrial Complex* (New York: The New Press, 2003).

22. Ibid.

23. Ibid.

24. "DPRK to Remove and Reprocess Spent Fuel from Yongbyon?" WISE-NIRS, *Nuclear Monitor,* no. 626, April 22, 2005.

25. Joseph Kahn, "North Korea Says It Will Abandon Nuclear Efforts," *New York Times,* September 19, 2005.

26. "DPRK to Remove and Reprocess Spent Fuel from Yongbyon?"

27. President Bush agreed to these proposals, according to several U.S. officials, because he was tied down in Iraq, dealing with the fallout from Hurricane Katrina, and heading toward another standoff over Iran's nuclear program. As the agreement unfolded, the Chinese exerted significant pressure upon the United States to either sign or take responsibility for a breakdown in the talks.

28. Joseph Kahn and David Sanger, "US-Korean Deal on Arms Leaves Key Points Open," *New York Times,* September 20, 2005.

29. "Weapons of Mass Destruction," Federation of American Scientists, http://www.fas.org/irp/threat/wmd.htm, August 17, 2000; and John Steinbach, "Israeli Weapons of Mass Destruction: A Threat to Peace," Center for Research on Globilization, March 2002.

30. "Egypt Proposes Nuclear-Free-Zone in Middle East," *Agence France Presse,* September 28, 2005.

31. Ibid.

32. Zia Mian and M.V. Ramana, "Feeding the Nuclear Fire," *Foreign Policy in Focus,* September 20, 2005.

33. Dr. S.P. Udaya Kumar, "India: The Energy Carrot and the China Stick," *Nuclear Monitor,* no. 633 (September 2, 2005).

34. Ibid.

35. Mian and Ramana, "Feeding the Nuclear Fire."

36. Praful Bidwai, "A Deplorable Nuclear Bargain," *Economic and Political Weekly,* July 30, 2005.

37. Kumar, "The Energy Carrot and the China Stick."

38. Ibid; and Steven S. Weisman, "US Allies and Congress 'Positive' about India Nuclear Deal," *New York Times,* July 20, 2005.

39. Mian and Ramana, "Feeding the Nuclear Fire."

40. Bidwai, "A Deplorable Nuclear Bargain."

41. Ibid.

42. Ibid.

43. Ibid.

44. Eric Margolis, "Fingers on the Button," *Toronto Sun,* August 14, 2005.

45. "Pakistan Nuclear Weapons: A Brief History of Pakistan's Nuclear

Program," Federation of American Scientists, http://www.fas.org/nuke/guide/ pakistan/nuke (accessed September 29, 2005).

46. Ibid.

47. Ibid

48. Ibid.

49. Ibid.

50. Ibid.

51. Ibid.

52. Ibid.

53. Salman Masood and David Rhode, "Pakistan Now Says Scientist Did Send Koreans Nuclear Gear," *New York Times*, August 25, 2005.

54. Ibid.

55. Caldicott, *The New Nuclear Danger*.

56. Ibid.

57. "World Leaders Shake Heads as Reforms to Check Nuclear Arms Spread Dumped," Agence France Presse, September 15, 2005.

58. Ibid.

59. Ibid.

60. Ibid.

9. Renewable Energy: The Answer

1. Timothy Egan, "Seeking Clean Fuel for a Nation, and a Rebirth for Small-Town Montana," *New York Times*, November 21, 2005.

2. Bruce Biewald, David White, Geoff Keith, and Tim Woolf, "A Responsible Energy Future, an Efficient Cleaner and Balanced Scenario for the US Electricity System," Synapse Energy Economics, Prepared for the National Association of State PIRGs Under Contract to the National Commission on Energy Policy, May 2005, US PIRG Reports.

3. Eric Martinot, "Renewables 2005: Global Status Report," REN 21 Renewable Energy Policy Network for the 21st Century, Paper Prepared for the Worldwatch Institute, www.ren21.net; and Amory Lovins, "Renewables to the Rescue: The Nuclear Write-Off," Rocky Mountains Institute, http://www.rmi .org/sitepages/pid171.php#E05-02.

4. John Nichols, "Enron: What Dick Cheney Knew," *The Nation*, April 15, 2002; Larry Klayman, interview by Bill Moyers, *NOW*, PBS, November 7, 2003; http://www.pbs.org/now/printable/transcript_klayman_print.html (accessed November 11, 2003); "Climate Scientists See Intimidation in Letter from House

Energy Chair," BushGreenwatch.org, July 13, 2005; http://www.bushgreen watch.org/; and Ross Gelbspan, "Katrina's Real Name," *Boston Globe,* August 30, 2005.

5. Amory Lovins, "Nuclear Power: Economics and Climate-Protection Potential," Rocky Mountains Institute, September 11, 2005.

6. If energy efficiency—lowering our *use* of electricy—is also factored in, "decentralized sources" of electricity generation in 2005 contributed ten times as much capacity per year as nuclear power.

7. Lovins, "Nuclear Power."

8. www.rmi.org/sitepages/pid171.php#E05-14.

9. If energy efficiency—lowering our *use* of electricy—is also factored in, "decentralized sources" of electricity generation in 2005 contributed ten times as much capacity per year as nuclear power.

10. Ibid.

11. In terms of global warming, however, it is important to note that electricity generation itself emits only 39% of the total U.S. CO_2; 61% comes from the *end uses* of energy, primarily transportation. See the next chapter for energy conservation ideas and for a guide to individual responsibility in the use of renewable electricity production.

12. John Deutch, et al., "The Future of Nuclear Power: An Interdisciplinary MIT Study," Cambridge, MA: MIT, 2003.

13. "Global Warming Versus Nuclear Power," editorial, *The New Scientist,* May 14, 2005.

14. David Stipp, "Katrina's Aftermath: The High Cost of Climate Change," *Fortune,* September 2, 2005.

15. Julian Borger, "U.S. States Bypass Bush to Tackle Greenhouse Gas Emissions," *The Guardian,* August 25, 2005.

16. Mark Townsend, "New U.S. Move to Spoil Climate Accord," *The Observer,* June 19, 2005.

17. Lovins, "Nuclear Power."

18. Matthew Wald, "Shifting Message: Energy Officials Announce Conservation," *New York Times,* October 4, 2005.

19. Paul Krugman, "Pig in a Jacket," *New York Times,* October 7, 2005.

20. Borger, "U.S. States Bypass Bush to Tackle Greenhouse Gas Emissions."

21. "U.S. Emissions in a Global Perspective," www.eia.doe.gov/oiaf/1605/ggrpt/pdf/chapter1.pdf (accessed August 26, 2005).

22. Lovins, "Nuclear Power."

23. Larry O'Hanlon, "A New Survey of Wind Power around the Globe Has Found There's Ample Energy for All Humanity Blowing around Us," *Discovery News*, May 24, 2005.

24. Ibid.

25. Ibid.

26. Wendy Williams, "The Danes Choose Wind Energy over Nuclear," Long Island Offshore Wind Initiative, http://www.lioffshorewindenergy.org/, June 7, 2005; and Harvey Wasserman, "Combines in the Sky, Farmer and Community-Owned Wind Power Set A New Green-Energy Trend in the US," *Renewable Energy World Magazine* (London), May–June, 2005.

27. Williams, "The Danes Choose Wind Energy over Nuclear."

28. Harvey Wasserman, e-mail communication, October 2005.

29. Ibid.

30. Ibid.

31. Ibid

32. Ibid.

33. Dan Juhl and Harvey Wasserman, "The True Free Market Choice: Let's Walk the Talk of Free Markets to Assess the Full Cost of Our Energy Options," Readers forum, *Solar Today*, March/April 2005.

34. Howard W. French, "In Search of a New Energy Source: China Rides the Wind," *New York Times*, July 26, 2005.

35. Ibid.

36. James Meek, "Back to the Future," *The Guardian*, October 4, 2005.

37. Ibid.

38. Gary Rivlin, "Green Tinge Is Attracting Seed Money to Ventures," *New York Times*, June 22, 2005.

39. Joseph Pereira, "Solar Power Heats Up," *Wall Street Journal*, June 2, 2005.

40. Ibid.

41. Craig D. Rose, "3 Billion Approved for Solar Rebates: Massive Program is Percira's Largest in U.S. History," *San Diego Union Tribune*, January 13, 2006.

42. Ibid.

43. Ibid.

44. Barry Rehfeld, "It's Getting Cheaper to Tap the Sun," *New York Times*, June 18, 2005.

45. Ibid.

46. "Global Warming Versus Nuclear Power," pp. 14–20.

47. Meek, "Back to the Future."

48. The Carbon Trust and D.T.I. Intermittency Literature Survey and Road Map, November 2003, Impact Study Annex 4, and "Variability of Wind Power and Other Renewables," *Management Options and Strategies*, IEA Publications, www.iea.org, June 2005.

49. Meek, "Back to the Future."

10. What Individuals Can Do: Energy Conservation and Efficiency

1. NationMaster.com, "Energy Usage Per Capita, Tonnes of Oil Equivalent," IEA, Energy Balances of OECD Countries 1999–2000, IEA, Paris, 2001.

2. Jerry Mander, *In the Absence of the Sacred*, San Francisco: Sierra Club Books, 1991.

3. Ibid.

4. Ross Gelbspan, "Katrina's Real Name," *Boston Globe*, August 30, 2005.

5. Bruce Biewald, David White, Geoff Keith, and Tim Woolf, "A Responsible Electricity Future: An Efficient, Cleaner and Balanced Scenario for the US Electricity System," Prepared for the National Association of State PIRGs Under Contract to the National Commission on Energy Policy, U.S. PIRG Reports, May 2005.

6. Ibid.

7. Ibid.

8. Ibid.

9. Ibid.

10. Ibid.

11. Nicholas Kristoff, "A Livable Shade of Green," *New York Times*, July 3, 2005.

12. Ibid.

13. Ibid.

14. Ibid.

15. Bruce Biewald, et al., "A Responsible Electricity Future."

16. Ibid.

17. Ibid.

18. Ibid.

19. Ibid.

20. Ibid.

21. David Adam, "Next Generation of Nuclear Reactors May Be Fast Tracked," *The Guardian*, January 21, 2006.

22. Michael Harrison and Michael McCarthy, "Plan for New Nuclear Programme Approaches Meltdown after Report," *The Independent*, March 7, 2006.

Index